UNITED STATES CRYPTOLOGIC HISTORY

Series V: The Early Postwar Period
Volume VI

The Quest for Cryptologic Centralization and the Establishment of NSA: 1940 - 1952

Thomas L. Burns

CENTER FOR CRYPTOLOGIC HISTORY

NATIONAL SECURITY AGENCY
2005

Table of Contents

Forewords

The Center for Cryptologic History (CCH) is proud to publish the first title under its own imprint, Thomas L. Burns's *The Origins of NSA*.*

In recent years, the NSA history program has published a number of volumes dealing with exciting and even controversial subjects: a new look at the Pearl Harbor attack, for example. Tom Burns's study of the creation of NSA is a different kind of history from the former. It is a masterfully researched and documented account of the evolution of a national SIGINT effort following World War II, beginning with the fragile trends toward unification of the military services as they sought to cope with a greatly changed environment following the war, and continuing through the unsatisfactory experience under the Armed Forces Security Agency. Mr. Burns also makes an especially important contribution by helping us to understand the role of the civilian agencies in forcing the creation of NSA and the bureaucratic infighting by which they were able to achieve that end.

At first glance, one might think that this organizational history would be far from "best seller" material. In fact, the opposite is the case. It is essential reading for the serious SIGINT professional, both civilian and military. Mr. Burns has identified most of the major themes which have contributed to the development of the institutions which characterize our profession: the struggle between centralized and decentralized control of SIGINT, interservice and interagency rivalries, budget problems, tactical versus national strategic requirements, the difficulties of mechanization of processes, and the rise of a strong bureaucracy. These factors, which we recognize as still powerful and in large measure still shaping operational and institutional development, are the same ones that brought about the birth of NSA.

The history staff would also like to acknowledge a debt owed to our predecessors, Dr. George F. Howe and his associates, who produced a manuscript entitled "The Narrative History of AFSA/NSA." Dr. Howe's study takes a different course from the present publication and is complementary to it, detailing the internal organization and operational activities of AFSA, and serves as an invaluable reference about that period. The Howe manuscript is available to interested researchers in the CCH.

It remains for each reader to take what Tom Burns has presented in the way of historical fact and correlate it to his/her experience. This exercise should prove most interesting and illuminating.

Henry F. Schorreck
NSA Historian
1990

*The Quest for Cryptologic Centralization and the Establishment of NSA: 1940-1952 is an unclassified version of *The Origins of NSA*, released for public review.

It may not be politically correct to say it, but the National Security Agency may be the last institutional legacy of the New Deal. NSA was established by President Harry S. Truman in late 1952, drawing on activities and organizations that traced from the 1930s and 1940s.

The United States had had vigorous communications intelligence (COMINT) activities in the American Civil War and World War I. After each of these conflicts, however, the organizations had been abolished and the activities ended.

Although the contribution of cryptology — both cryptanalysis and cryptography — was so far-reaching on the operations of World War II, there was little question that the activities would continue. However, the shape of the organizations to do it, their institutional home, and the level of effort were far from certain.

The post-WWII period in American politics was a bad time for fostering cryptology. The war had shown that it was necessary to use expensive machines to make and break codes or ciphers. Yet, this was a time of drastic budget cuts and personnel reductions in the government. For the major figures in the national security field, cryptology was a small but important issue; they were concerned primarily with the question of unification of the armed services and the possible creation of a central intelligence organization.

Thomas Burns clarifies the struggle of disparate forces concerned with cryptology, and shows clearly the intermediate steps taken before the creation of NSA. He deals forthrightly with the positive and negative aspects of the predecessor organizations, with their accomplishments and missed opportunities, and the factors that triggered the emergence of a new COMINT organization.

This is a story that in the past has been only partially glimpsed. Mr. Burns originally intended this as an in-house study. It is now available to help those interested in intelligence or cryptologic history to understand the forces that produced a major institution in the field.

For background, the reader might also want to consult Robert L. Benson's *A History of U.S. Communications Intelligence during World War II: Policy and Administration*, published by the Center for Cryptologic History in 1997.

David A. Hatch
NSA Historian
2005

Acknowledgments

Many people played a part in developing this expanded report on the origins of NSA. In particular, I want to extend thanks for the research help I received from the Office of Archives and Repository Services of the Telecommunications Organization and from my colleagues in the Center for Cryptologic History (CCH).

During the saga associated with the preparation of this volume, I became indebted to Henry F. Schorreck, NSA Historian, for his continuing confidence and support. In the latter stages of the review process, David W. Gaddy, Chief, CCH, and David A. Hatch, drawing on their broad experience and knowledge of the Agency , provided fresh perspective and comments. My appreciation goes to Arleone (Archie) M. Fisher for entering the draft manuscript into the word processor. Among the cycle editors, Gwen L. Cohen provided the initial editing of text, as well as the selection of photographs. Barry D. Carleen and his staff provided the editing actions associated with the final preparation of the manuscript. Special thanks are due to Jean M. Persinger and Donald K. Snyder for their editing contributions.

Finally, my appreciation goes to Gerald K. Haines, our former colleague, for his unfailing assistance in the shaping and organizing of the report, for his enthusiasm about the project and its scope, and for his helpful suggestions for sharpening and clarifying the text.

Thomas L. Burns

Introduction

The Struggle to Control a Unique Resource

More than sixty years have passed since the outbreak of World War II. During that war, a small number of organizations provided the total intelligence gathering activities of the United States government. Army and Navy authorities played a preeminent role in the production of this intelligence. Since 1945 a great number of organizational changes have occurred in the management and direction o f U.S. intelligence activities, and the intelligence community has greatly expanded. There is now a National Security Council (NSC), Central Intelligence Agency (CIA), Defense Intelligence Agency (DIA), National Foreign Intelligence Board (NFIB), and National Security Agency (NSA), as well as the military services, Department of State, Federal Bureau of Investigation (FBI), Department of Energy, Department of the Treasury, and Commerce Department. All are involved in intelligence activities, and all rely on or have access to communications intelligence (COMINT). COMINT is a unique, extremely valuable intelligence source. This study traces the evolution of the military structures from the early 1930s to the establishment of a unique agency to deal with COMINT–the National Security Agency–in 1952.

In the late 1930s, the major COMINT issue among the services related to the coverage of foreign diplomatic targets. Regardless of duplication, each service insisted on holding onto whatever diplomatic targets it could intercept. The realities of World War II, however, finally forced the services to work out an agreement on wartime cryptanalytic tasks. The Navy, because of its limited resources and its almost total preoccupation with Japanese and German naval traffic, ultimately softened its position and asked the Army to take over the entire diplomatic problem for the duration of the war. Based on an informal agreement by the Army and the Navy, the Army assumed responsibility for all targets in the diplomatic field, as well as its own commitments in the military field.

As late as 1942, however, the Army and Navy still resisted the introduction of any major changes in their relationship and sought to maintain their traditionally separate cryptanalytic roles. Each service worked independently and exclusively on its assigned cryptanalytic tasks, as was agreed upon previously, and later endorsed by President Franklin D. Roosevelt. The services not only continued to demonstrate little enthusiasm toward closer cooperation in COMINT matters, but maintained their traditional hostility towards proposals for merger, or even towards opening up new dialogue on operational problems. Consequently, their interaction on COMINT matters was minimal.

Nevertheless, out of the disaster at Pearl Harbor came persistent demands for the establishment of a truly centralized, permanent intelligence agency. As early as 1943, proposals for the establishment of a single United States intelligence agency became the routine topic for discussion in the various intelligence forums of the Joint Chiefs of Staff. At the same time, the military COMINT authorities foresaw their vulnerability to congressional criticism and future reductions in resources since they conducted their COMINT operations on a fractionated and sometimes duplicatory basis. Recognizing these threats to a continuation of their separate existence, the services, after two years of superficial coordination, established closer technical cooperation among their COMINT organizations.

During the war, the independent Army and Navy organizations accomplished a great number of spectacular intelligence successes in support of the Allied war effort. These included cryptanalytic breakthroughs against the communications of German submarines, German and Japanese armed forces, and the diplomatic communications of the Axis countries of both the European and Pacific theaters. The victory at Midway and the submarine war in the Atlantic are but two examples of how intelligence derived from enemy communications contributed to the success of the U.S. war effort. Ironically, these successes later became the measuring rod for criticism of the postwar military COMINT organizations.

By the end of World War II, many policy makers had a new respect for COMINT. However, there were also major questions concerning the management and control of this valuable resource. In 1951 President Truman established a presidential commission under the chairmanship of George A. Brownell to study the communications intelligence effort and to make recommendations concerning the management of the effort. From the Brownell Report grew the managerial foundation of the organization now known as the National Security Agency.

This study documents the origins of the National Security Agency. It is an attempt to set before the reader the "what happened" in terms of the issues and conflicts that led to Truman's decision to establish a centralized COMINT agency. It traces the evolution of the military COMINT organizations from the 1930s to the establishment of the National Security Agency on 4 November 1952.

While the lineal origins of the National Security Agency are clearly traceable to the military COMINT structures and represent a fairly simple audit trail of organizations, there is more to the origin of NSA than a mere chronology of organizations. The political struggles and operational considerations that led to the establishment of NSA are complex. The National Security Act of 1947, the expanding intelligence requirements of the growing intelligence community,

and the continuing controversy between the military and civilian agencies over the control of intelligence became prominent factors in the move to reorganize the nation's cryptologic structure.

This account seeks to highlight the main events, policies, and leaders of the early years. Its major emphasis is directed toward communications intelligence and its identification as a unique source of intelligence information. One theme persists throughout: the jurisdictional struggle between the military and civilian authorities over the control and direction of the COMINT resources of the United States. Special attention is also directed toward consumer relationships, intelligence directives, and consumer needs–particularly when those considerations may have influenced the shaping and formulation of the cryptologic structure.

The communications security (COMSEC) role of NSA is addressed only in the broad context of representing a basic responsibility of the new agency. The development of national COMSEC policies did not take place until after the establishment of NSA, which is outside the scope of this report. As directed by President Truman on 24 October 1952, the solutions to national COMSEC problems and the formulation of those solutions in directives became the responsibility of a special committee of the National Security Council for COMSEC matters. The beginnings of an expanded COMSEC role for NSA did not occur until the mid-1950s, following the issuance of a preliminary report (NSC 168) on 20 October 1953, which provided the basis for a later clarification of COMSEC roles and responsibilities within the government.

The study is organized basically in a chronological approach with chapters reflecting the prewar period, the war years, and the immediate postwar era. Major events or policy actions are reflected within this chronology. The early chapters address the evolution of the Army and Navy COMINT relationships from 1930 through the war years and later the establishment of a third cryptologic service, the Air Force Security Service (AFSS), in 1948. Next, emphasis is placed on the three-year period from 1946 to 1949, which marked the passage of the National Security Act of 1947 and the beginning of high-level efforts to centralize U.S. intelligence responsibilities. This section traces the organization of the COMINT structure as military authorities moved in the direction of a joint Army and Navy Communications Intelligence Board (ANCIB) and closer cooperation. This period of experimentation included the establishment of the Joint Army and Navy Operating Plan in 1946 and of the Armed Forces Security Agency (AFSA) in 1949. Both structures encountered great difficulties, with AFSA receiving continuing criticism from the consumer community for its performance during the Korean War. Finally, as the prologue to the establishment of NSA, there is an extensive discussion of the Brownell Committee, including the reasons for its establishment and the nature of its deliberations. The study concludes with an overall review of the organizational changes and a suggestion that struggle for control of this unique resource is far from over.

Chapter I

Early Army-Navy COMINT Relations, 1930-1945

During the 1930s and throughout World War II, the United States Army and the United States Navy dominated the U.S. COMINT effort. The Army and Navy COMINT organizations operated as totally autonomous organizations. They were fiercely independent, with little dialogue or cooperation taking place between them. Their working relationship represented a spirit of strong rivalry and competition, with overtones of mutual distrust. During the first two years of the war, the Army and Navy persisted in maintaining their totally separate cryptanalytic roles. Each worked independently and exclusively on its assigned cryptanalytic tasks, as approved earlier by the Joint Chiefs of Staff and President Franklin D. Roosevelt. Each service continued to oppose cooperation in COMINT matters. Both maintained a traditional hostility toward thoughts of merger, or even of opening up a dialogue with the other on cooperation.

Near the end of World War II, as the service COMINT organizations foresaw major reductions in their COMINT programs, their attitudes toward cooperation began to change. Moreover, as the pressures mounted for organizational change in the entire U.S. intelligence structure, the service COMINT authorities now initiated voluntary moves toward closer interservice cooperation, primarily as a self-preservation measure. In 1944, for example, the services expanded their cooperation on operational functions related to collection and cryptanalysis. The services also established the last joint forum for discussion of cryptologic matters, the Army-Navy Communication Intelligence Coordinating Committee (ANCICC). ANCICC, in turn, quickly evolved into the first overall COMINT policy board, the Army-Navy Communication Intelligence Board (ANCIB). As

further evidence of the broadening of the COMINT base. the Department of State accepted an ANCIB invitation to join the Board in December 1945. A civilian agency was now a part of the COMINT decision-making process.

As early as the World War I era, the U.S. Army and Navy COMINT organizations intercepted and processed foreign military and non-military communications for intelligence purposes. For all practical purposes, each functioned on a totally autonomous basis. Each service operated independently of the other, and each conducted its own intercept and exploitation activities. In this early period, intelligence requirements did not exist as we know them today. Generally. each service determined its own intercept targets and then, based on its own processing priorities, decrypted or translated whatever communications could be exploited. The Army and the Navy COMINT organizations disseminated the decrypts to the intelligence arms of their parent services, as well as to other governmental officials.

Except for a very restrained and limited exchange concerning cryptanalytic techniques, little cooperation or dialogue took place between the military COMINT organizations. Traditionally, each worked exclusively on those military and naval targets of direct interest to itself. Thus, the Army handled military radio stations and military messages, and the Navy handled naval radio stations and naval messages. The coverage of diplomatic targets, however, reflected a totally different story.

The coverage of diplomatic links always ranked as a top priority for both the Army and Navy, as it represented the only intelligence of real interest to non-military consumers, namely, the Department of State and the White House.[1] Recognizing the need for budgetary support from these influential customers, each service sought to retain a posture of maximum coverage on diplomatic targets. Consequently, the Army and Navy, operating under an unwritten and loose agreement, shared the responsibility for the intercept and processing of diplomatic traffic, with each service making its own determination concerning what diplomatic coverage it would undertake.[2]

Despite the increasingly apparent need for cooperation, neither service, because of strong mutual distrust, pressed very hard for a cooperative agreement. The nearest that the services came to concluding an agreement during the 1930s occurred in April 1933. The occasion was a planning conference of representatives of the War Plans and Training Section of the Army and representatives of the Code and Signal Section of the Navy.[3] The agenda for the conference was very broad, including items on both communications security and radio intelligence matters. The conferees reached a very limited informal agreement on a delineation of the areas of paramount interest to each service.[4] Although formal imple-

Rear Admiral Leigh Noyes Director of Naval Communications

mentation of the agreement never took place, the conference itself was a significant milestone. For the first time in the modern era, the services had agreed, at least in principle, on the need for a joint Army-Navy dialogue on COMINT matters.

From 1933 to 1940, little change took place in this relationship. Each service continued to go its own way, working generally on whatever traffic was available to it. In the fall of 1939, General Joseph A. Mauborgne, Chief Signal Officer, U.S. Army, and Rear Admiral Leigh Noyes, Director of Naval Communications, attempted an informal agreement concerning diplomatic traffic.[5] They agreed that diplomatic traffic would be divided between the two services on the basis of nationality. This agreement, like the one in 1933, however, was never implemented. The Army Signal Corps, on orders from its General Staff, worked on German, Italian, and Mexican diplomatic systems, thereby duplicating the Navy's efforts in this area. This effort completely nullified the earlier agreement negotiated by Mauborgne and Noyes.[6]

General Joseph A. Mauborgne Chief Signal Officer, U.S. Army

Colonel Spencer B. Akin

Commander Laurance F. Safford

By the summer of 1940, the war in Europe, coupled with the increasingly warlike posture of the Japanese in the Pacific, brought renewed pressures for closer Army-Navy cooperation. In addition, changes occurred in some foreign cryptographic systems that foretold the beginning of new technical challenges for both services. Despite the strong service antagonisms, the inevitability of closer cooperation and pooling of COMINT resources in some manner became apparent to many Army and Navy officials.

In mid-1940, a new round of formal Army-Navy discussions took place concerning "Coordination of Intercept and Decrypting Activities." The services established a Joint Army-Navy Committee, under the chairmanship of Colonel Spencer B. Akin and Commander Laurance F. Safford, to develop a method of dividing intercept traffic between them.[7]

The Army and Navy planners had no problem in reaching agreement on the division of responsibility for the coverage of counterpart targets. They simply opted for the status quo in the intercept coverage of military and naval targets. Thus, the Army retained the sole responsibility for the intercept and analysis of all foreign military traffic, and the Navy concentrated on the intercept and analysis of all foreign naval radio traffic.[8]

The discussions, however, failed to generate a solution to the issue of diplomatic coverage. Each service presented a number of proposals and counterproposals, but neither would yield any of its responsibility for coverage of diplomatic traffic.[9] The primary diplomatic targets under discussion at this time were German, Italian, Mexican, South American, Japanese, and Soviet.

Given the attitudes of the two services, there seemed little likelihood of achieving any agreements on diplomatic targets. The Army, having canceled the earlier 1939 understanding with the Navy, continued to work on German, Italian, and Mexican diplomatic systems, as well as on a number of machine problems of interest to the Navy. By 1940, the Navy, because of its heavy commitment to operational naval problems, stopped working on the German, Italian, and Mexican diplomatic targets. As a matter of principle, however, the Navy refused to concur on the exclusive assignment of these diplomatic targets to the Army on a permanent basis.[10] Another even more contentious item arose at the conference concerning the coverage of Japanese diplomatic traffic. Japan had become a prime intelligence target whose diplomatic communications were obviously of paramount interest and importance to each service – as well as to civilian U.S. officials. Neither service would relinquish any coverage of Japanese diplomatic communications.

In short, the joint conference resolved little. Since each service COMINT organization viewed its survival as being contingent upon the production of diplomatic intelligence, neither consented to giving up diplomatic coverage on a permanent basis. Colonel Akin and Commander Safford finally opted to refer the matter to their superiors – General Mauborgne and Admiral Noyes – for a decision on how to divide the Japanese COMINT problem.

As a last resort, Mauborgne, attempting a Solomonic approach, suggested that the Army and Navy simply alternate daily in their diplomatic coverage of certain functions such as decryption and translation duties. Adopting this suggestion as the way out of their dilemma, Mauborgne and Noyes informally concluded an agreement in August 1940, which became known as the "odd-even day" agreement. The agreement established the immediate prewar basis for the division of labor on all Japanese intercepts and delineated the responsibilities for decryption, translation, and reporting of Japanese diplomatic traffic. [11]

Under the terms of the agreement, the Army assumed responsibility for decoding and translating the intercepts of the Japanese diplomatic and consular service on the even days of the month. The Navy became responsible for translating the messages of the Japanese diplomatic and consular service on the odd days of the month. The agreement also included a restatement of the COMINT responsibilities for the intercept of Japanese military and naval traffic. The Army retained its responsibility for decoding and translating intercepts of the Japanese Army (including military attachés). The Navy continued to have the exclusive responsibility for the intercept and translation of the Japanese Navy targets (including military attachés). [12]

As a corollary to the informal odd-even arrangement, Mauborgne and Noyes ratified a supplemental technical agreement on 3 October 1940 concerning the division of intercept. [13] Colonel Akin and Commander Safford countersigned this agreement for the Army and Navy COMINT organizations. The agreement essentially represented a joint analysis of the existing intercept facilities and their capabilities. It reiterated the need for closer Army-Navy cooperation in order to provide better intercept coverage and to reduce duplication of effort. The report also reflected the considerable reliance placed at that time on the courier forwarding of traffic, both by air mail and surface transport, to achieve timeliness. [14] During the early part of World War II, the intercepted traffic was sent by sea, or by aircraft, and often arrived months late at its destination. This situation gradually changed as new radio teletype systems were installed.

The Mauborgne-Noyes odd-even verbal agreement remained in effect from 1 August 1940 until shortly after the Japanese attack on Pearl Harbor on 7 December 1941. [15] This odd-even arrangement proved to be fundamentally unsound. Cryptanalytically, alternating the responsibility for reporting greatly increased the risk of error, duplication, and omission. It also destroyed the element of continuity so critical to COMINT reporting and cryptanalysis. [16] Politically, however, the odd-even day arrangement accomplished a public relations function that was vitally important to the services. This arrangement divided the problem equally and permitted each to retain visibility with the White House and those officials who controlled the budget process.

Ten years later, Admiral Joseph N. Wenger defended the odd-even day arrangement. He indicated that in 1940 each service had all the available intercept in the Japanese diplomatic traffic and in some cases the means for breaking them. As a result, whenever an important message was read, "each service would immediately rush to the White House with a copy of the translation in an effort to impress the Chief Executive." According to Wenger, the awkwardness of this situation was the main reason for the adoption of the odd-even day arrangement as the only acceptable and workable solution for the services. Wenger conceded that the odd-even arrangement for processing traffic was a strange one, but in his view it was practical since traffic could be readily sorted according to the cryptographic date.[17] Wenger did not mention that it also achieved its main purpose, as each service remained visible to the White House.

The Wenger view represents the pragmatic view traditionally taken by the Army and Navy authorities at the time. Later assessments, however, differ in their treatment of the odd-even split, and are generally not as charitable. In a recent NSA cryptologic history, author Fred Parker presents a new perspective on the issue. While recognizing the Navy's limited resources, compounded by the primacy of the war in the Atlantic, he contends that the Navy misgauged the relative importance of Japanese diplomatic communications, and in the process lost valuable time in its pursuit of the more critical Japanese naval targets. He concludes that "had Navy cryptanalysts been ordered to concentrate on the Japanese naval messages rather than Japanese diplomatic traffic, the United States would have had a much clearer picture of the Japanese military buildup and, with the warning provided by these messages, might have avoided the disaster of Pearl Harbor.[18]

The attack on Pearl Harbor brought about increased activity in the conduct of U.S. intelligence activities. By the spring of 1942, a growing number of U.S. agencies began to conduct their own communications intelligence operations. Agencies now engaged in COMINT activities included the Department of State, Federal Bureau of Investigation (FBI), and Federal Communications Commission (FCC), as well as the Army and Navy.[19] This proliferation of COMINT activities became a matter of great concern to the military COMINT organizations. Because of security considerations, as well as the

scattering of scarce analytic resources, the Army and Navy sought to restrict sharply the number of U.S. agencies engaged in the cryptanalysis of foreign communications.

Turning to the Joint Chiefs of Staff, the military authorities requested a high-level decision limiting governmental activities in COMINT matters. Since the Joint Chiefs of Staff had the national responsibility for adjudicating issues related to intelligence, it represented the only forum available for defining U.S. jurisdictional responsibilities in the field of cryptanalysis. In its response, the Joint Intelligence Committee (JIC) of the JCS established a new Inter-Departmental Committee, entitled "Committee on Allocation of Cryptanalytical Activities." This committee, which had the task of surveying the entire field of cryptanalysis in the United States, included members from the Army, Navy, and FBI. It scheduled a conference for 30 June 1942.[20]

Leaving nothing to chance in their advance preparations, Army and Navy officials held a number of closed meetings prior to the meeting of the full committee. They sought primarily to resolve their long-standing disagreement on coverage of diplomatic targets. Five days before their meeting with the FBI, the Army and Navy succeeded in reaching an agreement on the division of COMINT responsibilities between their organizations.[21] The solution, urgently promoted backstage, resolved the nagging question of how to allocate service responsibility for diplomatic traffic. At the Navy's request, on 25 June 1942 the Army-Navy participants agreed to transfer the entire diplomatic problem to the Army for the duration of the war.

Many factors contributed to this decision. One related directly to a question of COMINT resources and capabilities. At the time, the war was primarily a naval war in both the Atlantic and Pacific theaters. As a result, the Navy, with its limited personnel resources, wanted to place its total emphasis on the naval problems. It recognized that its original ambitions for COMINT activities far exceeded its level of COMINT resources. For example, because of the restrictive Navy policy permitting only military personnel to work on COMINT-related matters, the Navy had a grand total of thirty-eight people assigned to diplomatic operations.[22] The Army, however, with a larger and predominantly civilian organization, was doing relatively little in military cryptanalysis. Since military traffic was virtually impossible to copy at long distances because of the low power used, the Army had very little to

between the Army and Navy on the diplomatic problem. In addition, both services, now sensitive to criticisms following the attack on Pearl Harbor, were anxious to forestall future charges about duplication of effort, wasted COMINT resources, and critical delays in the reporting of intelligence information.[24]

The new agreement concerning the transfer of diplomatic coverage also included guidelines governing the dissemination of COMINT from diplomatic sources to U.S. authorities.[25] Despite the

War Department Munitions Building

work on except diplomatic traffic.[23] As a result, the Army was able to assume exclusive responsibility for the diplomatic field without prejudicing its work on military targets.

A second factor, known to be of great concern to the Navy, was the planned relocation of the Army's COMINT facility from the old War Department Munitions Building on Constitution Avenue in Washington, D.C., to a site near Frederick, Maryland. Because of the distance to Frederick, the Navy viewed such a relocation as virtually ending the close daily collaboration

transfer of the basic responsibility for the diplomatic problem to the Army, the prior Army-Navy arrangements for the dissemination of diplomatic COMINT product remained in effect. The Army continued to supply the State Department with intelligence, and the Navy supplied the president with COMINT product. Following the 25 June 1942 agreement, the Army provided translations and decrypts to the Navy for delivery in the Navy Department and to President Roosevelt.

At the insistence of the Army, the 25 June 1942 agreement was a purely verbal arrangement between the officers in charge of the cryptanalytic sections. Commander John R. Redman, USN, represented the Navy, and Colonel Frank W. Bullock, USA, spoke for the Army.[26] The agreement later became known as the "Gentlemen's Agreement." Despite its informal nature, this understanding constituted a landmark in terms of Army-Navy collaboration in cryptanalysis. The earlier agreements were, in effect, little more than agreements "to talk" and generally resulted in no changes in the service roles. This agreement, however, became the first joint arrangement of any substance and the one that determined the shape and scope of a later wartime cooperation between the Army, Navy, and FBI.

When the full committee of Army, Navy, and FBI representatives convened on 30 June 1942, it simply accepted the earlier Army-Navy agreement and formally incorporated its provisions in a new document. The new document also addressed other issues that directly influenced the scope of U.S. cryptanalytic actions for the next few years. The agreement concluded that the conduct of cryptanalytic actions should be confined exclusively to the Army, Navy, and FBI, and it established the wartime policy governing the dissemination of the intelligence. In addition, the committee created a permanent standing committee to monitor the implementation of the

Commander John R. Redman, USN

agreement and to facilitate resolution of any problem areas. The formal Agreement of 30 June 1942 now became the official benchmark for the division of cryptanalytic responsibilities within the United States.[27]

Colonel Frank W. Bullock, USA

On 6 July 1942, the Joint Chiefs of Staff reported to President Roosevelt that such an agreement had been reached, and recommended that other U.S. agencies be excluded from the field.[28] On 8 July 1942, Roosevelt instructed Harold D. Smith, Director of the Budget, to issue instructions "discontinuing the cryptanalytic activities of the Federal Communications Commission, the Office of Strategic Services, the director of Censorship as well as other agencies having this character ."[29]

The presidential memorandum did not relate to or affect the division of responsibilities developed by the Army, Navy, and FBI in the 30 June meeting. It was always clear (at least to the military participants) that the 30 June 1942 agreement, as endorsed by the president, was a wartime arrangement made primarily to eliminate the FCC, OSS, and others from the cryptanalytic field, and to restrict the COMINT activities of the FBI.[30]

In implementing the agreements of June 1942, the Army assumed the Navy's previous responsibility for all cryptanalysis on other than naval problems and naval-related ciphers.[31] Thus, all foreign military traffic and all diplomatic communications fell to the Army. The Navy acquired responsibility for enemy naval traffic, enemy naval air and weather systems, and through its wartime control of the Coast Guard, surveillance of clandestine communications. The conference concluded that there was sufficient clandestine material to occupy both the FBI and the Navy (Coast Guard) with reference to Western Hemisphere clandestine work since both were engaged in it and had a vital interest in the

tion of new and greatly expanded operational missions, each service initiated search actions for the acquisition of new sites to house their already overcrowded facilities. Within one year, the services accomplished major relocations and expansions of their operations facilities within the Washington, D.C., area.

The site ultimately selected by the Army came to the attention of the authorities quite by chance in the spring of 1942. On returning from an inspection of the proposed site for a monitoring station at Vint Hill Farms, near Warrenton, Virginia, the Search Team, among whom was Major Harold G. Hayes, Executive Officer, Signal

Aerial view of Arlington Hall Station

results. For other than the Western Hemisphere, the Navy (Coast Guard) acquired exclusive responsibility for international clandestine communications. The FBI in addition to sharing the responsibility with the Navy for clandestine targets in the Western Hemisphere, worked domestic voice broadcasts and domestic criminal actions.[32] This overall division of cryptanalytic effort proved to be an effective wartime arrangement.

Coincidental with the negotiations over the allocation of cryptanalytic targets and in anticipa-

Intelligence Service (SIS), happened to notice the impressive grounds and facilities of the Arlington Hall Junior College at 4000 Lee Boulevard, Arlington. Almost immediately, the Army sought to acquire the property, which was then in receivership. The property, as it turned out, was not on the governmental list for possible purchase, nor was it on the market at the time. The Army, however, sought to acquire it through a straight purchase arrangement, but failed to reach agreement with the seller on the price. Arlington Hall Junior College officials valued the buildings and grounds (approximately ninety-six acres) at $840,000 while the Army appraised the property at $600,000. Following litigation actions and condemnation of the property under the War Powers Act, the court established the final price at $650,000. The SIS took official possession of the property on 14 June 1942. By the summer of 1942, the Army's Signal Intelligence Service organization completed the move from the Munitions Building on Constitution Avenue in Washington, D.C., to its new location, now called Arlington Hall Station.[33]

The Army then began a major building program to accommodate the wartime expansion of personnel and equipment. The building pro-

to move into the new quarters as the spaces became ready for occupancy. By November 1942, however, the SIS announced a further expansion of its civilian personnel to a total of 3,683 employees. Consequently, the Army broke ground on 4 December 1942 for the construction of a second operations building similar to Operations A Building, but somewhat smaller in size. The new Operations B Building became fully occupied by 1 May 1943.[34]

The same situation applied to the Navy – namely, a pressing need for additional personnel, space, and security in order to meet the increased operational requirements of war. In 1942, Navy

Aerial view of Navy Headquarters at Nebraska Avenue, March 1949

gram provided for the construction of temporary buildings, without air conditioning or other refinements. The initial expansion included a rehabilitation of the main school building and the construction of barracks for enlisted men and operations buildings. In September 1942 the Army started construction of the new barracks and broke ground for the construction of a new operations building. Operations A Building, 607 feet long by 239 feet wide, provided approximately 240,000 square feet of floor space, and was designed to accommodate 2,200 personnel. Within two months, operational elements started

planning called for a major expansion of its COMINT unit, the Supplemental Branch (OP-20-G) located in the Navy Department on Constitution Avenue in Washington. The Navy, preferring to stay in Washington, acquired the site of the Mount Vernon Seminary at 3801 Nebraska Avenue. On 7 February 1943 OP-20-G moved from the Navy Department to its new site of thirty-five acres, now called the Communications Supplementary Annex, Washington (CSAW). Commensurate with the size of the property, which was considerably smaller than Arlington Hall Station, the Navy undertook a building program to meet its particular needs. Unlike the Army, however, the Navy tended to construct permanent buildings rather than temporary structures. The expansion of the CSAW site included modifications of existing structures, construction of new support facilities, and construction of a major new building.[35]

During the first two years of the war, the services continued to expand their COMINT resources, both in Washington and overseas. Despite the proximity of their COMINT headquarters and the working agreement, each service remained aloof and zealously guarded its own operations. Each worked independently of the other. By the end of 1943, however, with the end of the war in sight, the COMINT authorities in both services foresaw that the survival of their COMINT operations would be in jeopardy if they persisted in maintaining totally independent operations. This factor became the main catalyst in developing closer cooperation.

The year 1944 marked the beginning of a new period in Army-Navy collaboration in cryptanalysis. During 1944 the Army and Navy completed a number of supplemental agreements, all of which reflected logical extensions or clarifications of the earlier 1942 agreement and which moved in the direction of establishing closer coordination. On 19 January 1944, for example, a joint agreement signed by General George C. Marshall and Admiral Ernest J. King promulgated the "Joint Army-Navy Regulations for the Dissemination and Use of Communications Intelligence Concerning Weather." The agreement addressed the special nature and perishability of Japanese weather intelligence. Heretofore, the services traditionally handled weather intelligence as a special category of intelligence, with each having totally separate rules to govern its classification, handling, and dissemination. The King-Marshall agreement changed this by establishing new uniform security regulations to govern all U.S. services in their handling of Japanese Special

Weather Intelligence (SWI).[36] On 7 April 1944 an additional Army-Navy agreement defined the basic allocation of cryptanalytic tasks against Japanese weather systems. This second agreement included specifics on the realignment of cryptanalytic tasks on the weather problem, a new policy statement authorizing a complete exchange of all information concerning weather systems, and new guidelines governing the exchange of weather intelligence.[37]

The Army-Navy authorities also completed two additional policy agreements in 1944. On 4 February 1944 Marshall and King issued a "Joint Army-Navy Agreement for the Exchange of Communications Intelligence." This agreement, applicable only to the Washington area and only to the Japanese problem, provided for the first exchange of liaison officers between Army and Navy communications intelligence organizations. In addition to their liaison role, the officers were to "have access to the . . . intelligence files and records" in the Army (SSA) and Navy (OP-20-G) COMINT organizations.[38] (As the Army's Signal Intelligence Service evolved, it became the Signal Intelligence Division (1942); the Signal Security Service (1942); the Signal Security Agency (1943); and the Army Security Agency (1945). The second agreement, which formalized a long-standing working arrangement, concerned the sharing of communications circuits. During the early years of the war, the Army permitted the Navy to use the Army's communications circuits to Australia. In response to a Navy request, on 15 June 1944 Major General Harry C. Ingles, Chief Signal Officer, agreed to the "continued and perhaps increased movement of Navy traffic over the channels of Army Communications Service, extending between the United States and Australia."[39]

Despite the progress taking place in the course of Army-Navy cooperation in COMINT matters, the June 1942 agreement remained the dominant and most important component governing their intelligence relationships. While the services did agree to a minimal expansion of intelligence arrangements existing on the periphery of their basic dealings, neither

*Major General Harry C. Ingles,
Chief Signal Officer*

sought to amend the earlier agreement. As the war progressed, the cryptologic services continued to concentrate on the targets previously allocated to them. The Army processed the foreign diplomatic communications, while the Army and Navy targeted their efforts, on a counterpart basis, against the military and naval communications of Japan and Germany. This breakout of cryptanalytic tasks proved to be conceptually sound and completely acceptable to each service.

Also, the communications security practices of foreign countries reinforced the U.S. decision concerning the division of the intelligence effort between the Army and Navy. Developments during World War II indicated that there existed no centralized control within Japan and Germany over the development of their cryptosystems. The systems employed by the Japanese and German users were designed separately by their foreign offices and armed services. Having worked independently for so many years, this fractionation of foreign responsibility for crypto-development worked to the advantage of the Army and Navy in their technical efforts to exploit enemy communi-

cations. Moreover, the organizational design of the U.S. cryptologic structure facilitated the targeting of enemy communications on a totally decentralized basis by existing organizations – and without the requirement for establishing a unified center for the analysis of communications.[40]

Devoting extensive resources and talent to their missions, the Army and Navy COMINT organizations accomplished some remarkable exploitation of enemy communications during the war. The Army enjoyed extraordinary successes against foreign Japanese diplomatic traffic enciphered in an electro-mechanical system known as the "Red" and subsequently as the "Purple" machine. The Army also exploited the Japanese Water Transport Code, which it broke in April 1943, as well as other military systems.[41] In its exploitation of the Water Transport Code, which utilized a Japanese Army cryptographic system, the Army provided valuable advance warning about the movements of Japanese merchant shipping that operated in support of enemy ground forces and helped eliminate nearly ninety-eight percent of the Japanese merchant fleet by the end of the war. Similarly, the Navy enjoyed its share of spectacular successes against Japanese and German naval communications. Navy COMINT provided the breakthrough to JN-25, the major Japanese fleet system, that helped the United States win the battles of Coral Sea and Midway, the turning points in the Pacific War.[42] The Navy also achieved a number of other critical breakthroughs in the decoding of Japanese convoy messages, as well as German communications concerning the movements of submarines in the Atlantic.[43]

While their cryptanalytic accomplishments were impressive, both services recognized that their joint efforts were far from ideal. The war compelled them to develop closer COMINT relationships with each other, but there still persisted a highly competitive and frequently hostile relationship. Each service cooperated with the other to the extent agreed upon – but there was little evidence of enthusiasm or voluntary efforts to go beyond the formal arrangement. A spirit of competition rather than coordination continued throughout the early years of World War II. In such an atmosphere, the intensity of competing interests tended to create unnecessary difficulties for each organization. Recurring problems, such as the recruitment of suitable personnel or the procurement of highly complex and unique cryptanalytic machinery, were often complicated by the competition of both services for the same items.[44] Thus, limited coordination and the absence of free and open dialogue between the services on day-to-day operational relationships meant that the Army and Navy were often working at cross-purposes.

Even in such a sensitive area as foreign relationships, each COMINT service demonstrated a predisposition to act completely independently. For example, the Army and Navy persisted in establishing their own technical agreements with their British counterparts, but without coordination or dialogue with the other U.S. service. These agreements frequently conflicted, usually with respect to the amount and kinds of intelligence information to be exchanged. Because of these diverse agreements, a potential for serious damage to American intelligence interests always existed.[45]

Similarly, on U.S. intelligence matters, each service operated with little consideration for the parallel activities and interests of the other. A policy of non-coordination seemed to prevail, which applied particularly to the relationships of the intelligence services with each other and with the Office of Strategic Services (OSS) and the FBI. Lacking any central authority for intelligence activities, the services for the most part had free rein in their operations. An internal FBI memorandum in 1939 to FBI director J. Edgar Hoover testifies to the disunity in noting that "another feud had broken out between the Army and Navy Intelligence sections" because the Army Military Intelligence Division (G-2) had approved the request for representatives of the Japanese Army to examine certain plants and factories in the United States after the Navy had turned down the request.[46] This pattern of independent and autonomous operations by the intelligence services continued throughout the war years.

Since no national intelligence structure existed, the Joint Chiefs of Staff served as the primary U.S. mechanism to govern U.S. intelligence activities during World War II. The need for establishing a coordinating committee composed of representation from the various departments and agencies was recognized early in the war, resulting in the creation of the Joint Intelligence Committee (JIC) under the Joint Chiefs of Staff. At its first meeting in March 1942, the membership consisted of the intelligence chiefs of the Army and Navy, and one representative each from the Department of State, the Board of Economic Warfare (later the Foreign Economic Administration), and the Coordinator of Information (subsequently the Office of Strategic Services). The intelligence chief of the Army Air Corps was added in 1943.[47]

As its principal function, the JIC provided intelligence estimates of enemy capabilities for use in developing strategic war plans for the JCS. In addition, the JIC provided advice and assistance to the JCS on other intelligence matters and also served as a coordinator of intelligence operations conducted by the member agencies. The JIC, with its many subcommittees, provided the primary forum for community discussion of intelligence reports, estimates, requirements, and related topics. During the first two years of the war, it was within the JIC that the first joint producer and consumer discussions concerning COMINT matters took place. The topics included such recurring items as possible ways to improve COMINT product, COMINT dissemination procedures, and the matter of cooperation between the COMINT organizations.[48]

As early as 1942, however, it became evident that COMINT agencies were making independent decisions concerning requirements and the priorities of intercept, cryptanalysis, and reporting. While the consumer representatives such as State and OSS may have been uneasy about this situation, they were unable to change things because of their lack of influence in directing the overall COMINT structure. It was basically controlled by the U.S. military. Toward the end of the war, the question of how to influence and guide the collection and reporting priorities of the U.S. COMINT structure surfaced as a fundamental issue within the U.S. intelligence community. Nothing really changed at the time, however, and the ramifications of this unresolved issue extended well into the postwar period.[49]

In much the same vein, recognizing the magnitude of the intelligence picture, and seeking to benefit from "lessons learned" during the war, the Joint Intelligence Staff of the Joint Chiefs of Staff started to explore the concept of establishing a central intelligence organization for the United States for the postwar period. Brigadier General William J. Donovan, head of the Office of Strategic Services, became a major catalyst for these discussions.[50] In late 1944 Donovan, functioning as a member of the Joint Intelligence Committee of the JCS, presented his first proposal for the establishment of a central intelligence agency. Among other things, Donovan's proposal recommended the establishment of a national intelligence authority and a central intelligence agency. While Donovan's proposal generated much discussion in the JCS Committee structure, it never went beyond the proposal stage during the war and remained within the JCS structure.[51]

When discussing COMINT activities, the various JCS committees always emphasized the need for much closer cooperation by the COMINT producers. This consideration, coupled with the recurring proposals for centralization of intelligence activities, brought new fears to the COMINT service organizations, however. The

COMINT authorities in each service recognized that a disunited COMINT structure would be more vulnerable to a takeover in the event centralization of intelligence actually was forced upon them.[52]

Yet another consideration influenced the thinking of the COMINT hierarchy. Recalling an earlier parallel from World War I, both Army and Navy policy makers became apprehensive about the effect of demobilization on their COMINT organizations. They were concerned lest the situation that had occurred at the end of World War I might happen again – namely, dwindling appropriations and the inability to provide for future COMINT needs. In looking at the pattern following World War I, one Navy study concluded that "lost opportunities and neglect, which was the fate of all American military and naval enterprises in the postwar era, was suffered [sic] by United States Army and Navy Communication Intelligence organizations."[53] No one in the Army or the Navy wanted a repeat of the World War I experience.

Moving in the direction of still greater cooperation, on 18 April 1944 the services set up an unofficial working committee known as the Army-Navy Radio Intelligence Coordinating Committee. Members of the committee were Colonel Carter W. Clarke and Colonel W. Preston Corderman for the Army; and Captain Philip R. Kinney, Captain Henri H. Smith-Hutton, and Commander Joseph N. Wenger for the Navy. The committee's mission involved policy, planning, and technical matters. It met monthly, and, in general, worked on postwar plans, coordination of future operating plans in the Pacific, and coordination of relationships and agreements with allied radio intelligence activities. Initially, the committee had no formal organization and little official power.[54] Following its first two meetings (18 April and 19 May 1944), the committee changed its name to Army-Navy Communication Intelligence Coordinating Committee (ANCICC) to reflect the increasing usage of the term "communications intelligence" in place of "radio intelligence."[55]

The establishment of ANCICC represented a significant step forward in the area of service cooperation. There now existed a forum, albeit informal and with a limited charter, empowered to consider a broad range of COMINT problems. On controversial or critical issues, ANCICC lacked the authority to make decisions. COMINT officials from both services, such as Carter Clarke, Preston Corderman, and Joseph Wenger, recognized the obvious need for another, higher level board, with broader authority, to discuss COMINT problems independently of other forms of intelligence. Each service, therefore, agreed to study the possibility of establishing a higher level military board to govern COMINT matters.[56]

In less than a year, the services succeeded in establishing such a policy board. In an exchange of letters in 1945, Admiral Ernest J. King, Chief of Naval Operations and Commander in Chief, U.S. Fleet (COMINCH), and General George C. Marshall, Chief of Staff, U.S. Army, agreed in principle to the establishment of an Army-Navy Intelligence Board. Based on national intelligence interests, they considered it imperative that the Army and Navy intelligence organizations work more closely together on an interdepartmental and permanent basis.[57]

On 10 March 1945 Marshall and King cosigned a Joint Memorandum to the Assistant Chief of Staff (G-2), to the Commanding General, Signal Security Agency, to the Director of Naval Intelligence, and to the Director of Naval Communications, that formally established the Army-Navy Communication Intelligence Board (ANCIB). The Marshall-King memorandum defined the authorities and responsibilities of the new board, and redesignated the informal ANCI-CC as an official working committee of ANCIB.[58]

Because of security considerations, Marshall and King insisted that ANCIB function outside the framework of the Joint Chiefs of Staff and report directly to them. Their major concern about security was the exposure of sensitive COMINT information via the multilayered correspondence channels of the JCS. The placement of ANCIB within the JCS structure would have required the automatic routing of all papers and reports through the JCS Secretariat, thereby exposing ULTRA intelligence to personnel not considered as having the "need-to-know."[59]

According to its charter, ANCIB was established primarily to avoid duplication of effort in COMINT matters and to ensure a full exchange of technical information and intelligence between the services. However, it also included a self-restricting provision that required unanimity of agreement on issues requiring a decision by the board. This rule enabled the military COMINT structures to appear to coordinate operations on a voluntary basis without, in fact, yielding any of their independence. By simply exercising its veto power, a service could prevent the implementation of any controversial proposal. In later years, the rule of unanimity developed into a major problem for the entire intelligence community. Nevertheless, the Marshall-King agreement represented a significant milestone in service cooperation–the establishment of the first interdepartmental board devoted solely to COMINT matters. With the establishment of a joint Army-Navy board, the services created their own self-governing mechanism to administer their COMINT effort. When discussing the merits of establishing a new Army-Navy Communications Intelligence Board, Rear Admiral Joseph R. Redman presented very bluntly some of the fears and concerns of the military services. He stated in a letter to Vice Admiral Richard S. Edwards, Chief of Staff, Office of Naval Operations:

> *... The public is acutely conscious of the lack of unified direction. . . in American intelligence activities. The supposedly secret plan of the OSS for coordination of all these activities is widely known... In addition, there seems to be little doubt that other civilian agencies will insist on a reorganization of American*

intelligence activities. It is important that the Army and Navy take progressive steps. . . to ensure that their legitimate interests in communication intelligence are not jeopardized by the encroachment of other agencies. The formal establishment of an Army-Navy Communications Intelligence Board will ensure that communication intelligence, the most important source of operational intelligence, will be discussed independently of other forms of intelligence. . . . Finally, informed observers will have some assurance that nothing has been left undone to ensure that another Pearl Harbor will not occur.[60]

Whatever motivations may have contributed to the establishment of ANCIB, the new board became a powerful joint institutional force in the adjudication of COMINT matters, both at the policy and operational levels. The establishment of ANCIB did not diminish the competition between the Army and Navy COMINT organizations, however. Despite the new ANCIB, the services were more determined than ever to preserve their separate COMINT organizations. They viewed ANCIB as a valuable joint mechanism that would assist in sheltering their COMINT activities from external scrutiny and in the resolution of joint problem areas. But they also foresaw that the structure of the new board would permit each service to remain totally independent.[61]

In the closing days of the war, the services seemed to be driven by two compelling and overlapping objectives. First, they desired to find a way to formalize their joint day-to-day relationships as an initial step toward protecting the existing COMINT resources of the Army and Navy from drastic budget cuts. Second, they believed some way had to be found to continue the progress achieved during the war in conducting COMINT collaboration with Great Britain.[62] This wartime liaison with the British in COMINT proved to be highly beneficial to each country, as it permitted a sharing not only of cryptanalytic techniques but also of cryptanalytic successes.

The establishment of ANCIB marked the first step toward accomplishing the first objective. The Army and Navy now had an interdepartmental forum for joint discussion of COMINT matters. Accomplishing the second objective required considerable internal joint discussions, as well as a new round of external negotiations with British officials.

The origins of the highly secret U.S.-U.K. COMINT cooperation stemmed from the outbreak of war in Europe in 1939 and the German blitzkrieg through sections of Western Europe. By the summer of 1940, Great Britain, under heavy siege by German forces, intensified its efforts to acquire military assistance from the then neutral United States. President Roosevelt, at least in the early days of the war, sought to achieve a partial posture of neutrality for the United States, but it was evident that he personally favored a policy of "all-out aid" to Great Britain. When Winston Churchill became the prime minister of the United Kingdom in May 1940, Roosevelt and Churchill quickly established a direct and personal communications channel on matters related to the war. This extraordinarily close relationship of the two leaders reinforced the concept of a strong British-U.S. alliance, and influenced many of the joint military decisions during the course of the war.[63]

In 1940, following a number of high-level conferences in London and Washington, the two governments concluded "a general, though secret agreement. . . for a full exchange of military information."[64] Following this very broad agreement, the United Kingdom and United States representatives began very limited exploratory talks in August 1940 concerning the establishment of a cryptanalytic exchange between the two nations. These discussions marked the beginning of a cryptanalytic exchange, but one that functioned on a very limited and cautious basis and that took place at a service-to-service level.[65] Close cooperation did not actually begin until February 1941. Each U.S. service, from the outset, worked independently in developing its own agreements or understandings with its British counterpart and seldom told the other of its accomplishments.

In July 1942, basically unaware of the service competition in this field, Prime Minister Churchill brought up the subject by informing Roosevelt that the British and American naval "cipher experts" were in close touch but that a similar interchange apparently did not exist between the two armies.[66] Roosevelt asked Marshall to take this up with Field Marshal Sir John Dill, British Ministry Office Liaison Officer in Washington.[67] In a response to Marshall's request for information, on 9 July 1942 Major General George V. Strong, Assistant Chief of Staff, G-2, stated that an interchange of cryptanalytic information between the British and American armies had been taking place for over a year and that it appeared to be satisfactory to both sides.[68] Strong further stated that if the Navy exchange of cryptanalytic information with the United Kingdom seemed to be more advanced, it was simply because coordination between the two had been necessary for a much longer time.

In 1943, however, the U.S. Army and British authorities completed a formal agreement concerning collaboration on their major military COMINT targets. Under the agreement, the U.S. Army assumed as a main responsibility the reading of Japanese military and air traffic. The British Government Code and Cypher School (G.C. & C.S.) assumed a parallel responsibility for a cryptanalytic effort against German and Italian military and air traffic. The agreement provided for complete interchange of technical data and special intelligence from the sources covered, and for dissemination of such intelligence to all field commanders through special channels. On 10 June 1943, Major General Strong signed the agreement for the U.S. War Department, and

Edward W. Travis, Deputy Director, G.C. & C.S., signed for the British."[69]

Thus, during the war years the Army and Navy followed the established policy of working independently with the British, with each U.S. service having separate agreements or understandings with its British counterpart. In general, because of mutual distrust, each consistently failed – or refused – to inform the other of the existence or nature of their agreement with the United Kingdom.

Franklin D. Roosevelt and Winston Churchill established a direct channel of personal communications in 1940, reinforcing strong ties between the United States and Great Britain.

Major General George V.
Strong
Assistant Chief of Staff, G-2

Edward W. Travis
Deputy Director, G.C. & C.S.

The first clear indication that the services were beginning to be more open with each other concerning their foreign COMINT arrangements occurred during 1944. This change of attitude came about, in part, because of the establishment of the new Army-Navy Communications Intelligence Coordinating Committee; a general acceptance of the need for tighter control of foreign agreements on COMINT matters; and the likelihood of continuing U.S. collaboration with the British.[70] Using the informal forum provided by ANCICC, each service began to reveal the specifics of its agreements with foreign nations, especially their COMINT relationships with Great Britain. It was small progress and did not undo immediately the independent agreements made earlier by each U.S. service with foreign organizations. Nevertheless, it was progress.

In exploring the possibility of establishing postwar collaboration with the United Kingdom, many alarming reports emerged about the earlier lack of Army-Navy coordination and the existence of overlapping agreements with the United Kingdom. For example, the Navy noted, "the lack of coordination between the Army and the Navy was strikingly demonstrated by an Army-British agreement which was made during the war without the concurrence of the Navy, even though it directly affected the air material in which the

Navy had a vital interest. It also provided for a complete exchange between the Army and the British of all technical material, although the Navy had an agreement to make only a limited exchange with the British."[71] The Navy cited similar problems in some of its COMINT relationships with U.S. consumers. It noted that both services experienced similar difficulties stemming from their unilateral dealings with the FBI and OSS. It was not until the creation of the informal ANCICC that the Army and Navy achieved a united front in dealing with these agencies.[72] By the end of the war, Army and Navy officials came to realize that COMINT agreements with foreign governments or other domestic agencies could no longer be determined on an ad hoc basis by each service.

At the same time, ANCIB undertook its own efforts to strengthen the U.S. COMINT structure. As a part of this effort, ANCIB sought to find a way to continue U.S.-U.K. collaboration in COMINT, and to establish itself as the sole U.S. spokesman for the conduct of policy negotiations with all foreign countries on COMINT matters.[73]

The board saw British-United States cooperation as the key. By early 1945, as the primary wartime targets began to dry up, Great Britain and the United States began a redirection of their

COMINT efforts. At that time, there emerged a dominant view among the allied nations that the Soviet Union was a hostile and expansive power with whom good relations seemed highly unlikely, at least for the immediate future. Since both nations recognized the mutual benefits of their earlier collaborative efforts, they agreed to investigate the feasibility of establishing some form of postwar collaboration on the Soviet problem that heretofore had received minuscule attention. Motivated in large part by sensitivity and security considerations, and seeking to avoid a repetition of the many separate wartime agreements with each other, representatives of both nations agreed that a new set of ground rules would be necessary for their next round of collaboration. Both authorities agreed to establish mechanisms within their own countries to bring about a greater degree of centralized control of their COMINT resources. This move to establish a new cycle of British-U.S. collaboration also meant that the U.S. services would have to be more open with each other about their COMINT programs and successes.

Within the U.S. intelligence structure, the Army and Navy now endorsed the concept of centralized control to govern their foreign COMINT relationships. In the negotiation process, the services agreed to the establishment of well-defined policies and procedures to govern the conduct of United States COMINT liaison with the British COMINT authorities. Under the new concept, all U.S. foreign liaison with Great Britain on certain problems would take place under the auspices of the United States policy board (ANCIB/ANCICC) rather than individually by each service.

Five years after the initial U.S.-U.K. collaboration in COMINT, the two nations began a new chapter in their cooperation in COMINT matters. Following several months of technical discussions, both in London and Washington, representatives of the London Signals Intelligence Board (LSIB) and the Army-Navy Communications

Intelligence Board on 15 August 1945 informally approved the concept of establishing U.S.-U.K, cooperation on certain problems.[74] This unwritten agreement was predicated on an understanding arrived at by Group Captain Eric Jones, RAF, and Rear Admiral Hewlett Thebaud, Chairman of ANCIB. The informal understanding identified LSIB and ANCIB as the respective governmental authorities for all COMINT negotiations and outlined in general terms the framework and procedures to govern the new working partnership.

This growing collaboration soon resulted in a broad exchange of operational materials between the COMINT centers of both nations, and in the establishment of reciprocal Joint Liaison Units stationed in London and Washington. These liaison units evolved into the liaison mechanisms that exist today, the Senior U.S. Liaison Officer, London (SUSLO), and the senior U.K. Liaison Officer, Washington (SUKLO).[75]

In the implementation of this increased cooperation, ANCICC established the Army and Navy COMINT organizations as its focal points, to serve on a rotating basis for the conduct of liaison with Great Britain. Initially, an Army officer represented ANCICC in London, assisted by a Navy officer. Similarly, a naval officer, assisted by an Army officer, represented ANCICC in Washington. This detail rotated every six months, so that first one service represented ANCICC as senior liaison officer, followed by a member of the other service. This system worked very well. It served to keep each service in the forefront on operational and policy matters while at the same time providing a new degree of centralized control over COMINT activities under the aegis of ANCIB. It also helped to prevent the United Kingdom from playing one service off against the other, as had occurred frequently during World War II.[76]

These arrangements became the springboard for further U.S.-U.K. negotiations to consider the establishment of even broader collaboration for the postwar period. With this objective in mind, the COMINT authorities brought the matter of U.S.-U.K. collaboration to the attention of President Truman through the State-War-Navy Coordinating Committee (SWNCC). In 1945, in a joint memorandum to Truman, Acting Secretary of State Dean Acheson, Secretary of War Henry L. Stimson, and Secretary of the Navy James V. Forrestal recommended the continuation of collaboration between the United States Army and Navy and the British in the field of communications intelligence.[77] On 12 September 1945, Truman concurred. He authorized the Army and Navy ". . . to continue collaboration in the field of communication intelligence between the United States Army and Navy, to extend, modify. or discontinue this collaboration as determined to be in the best interests of the United States."[78] Based on this presidential authorization, the Army and Navy immediately initiated U.S.-U.K. discussions through ANCIB to explore expanded postwar collaboration in COMINT .[79]

As ANCIB pursued its objectives, however, a new COMINT unit, outside the military structure, appeared in the U.S. intelligence community: the Department of State unilaterally established its own unit to exploit COMINT. Because of the desire to bring all the COMINT activities of the United States under the control of ANCIB, ANCIB officials agreed to seek the expansion of its membership to include State.

On 13 December 1945, ANCIB forwarded its proposal for expansion of the board to General Eisenhower and to Admiral King for approval. They approved the recommendation, and the Department of State accepted membership on ANCIB, effective 20 December 1945. ANCIB and its working committee became the State-Army-Navy Communication Intelligence Board (STANCIB) and the State-Army-Navy Communication Intelligence Coordinating Committee (STANCICC).[80] Alfred McCormack, special assistant to the secretary of state, became the first State Department member of STANCIB.[81] A civilian agency was now an official part of the United States COMINT structure.

In summary, by the end of the war. the United States COMINT services had reason to be proud of their accomplishments. They had achieved spectacular COMINT successes against the military and diplomatic communications of Germany and Japan. To achieve a greater degree of efficiency and to avoid costly duplication, they had set up their own self-governing mechanisms—a policy board (ANCIB) and a working level committee (ANCICC). Despite all their efforts, however, they still basically functioned as independent units in the COMINT arena.

These successes notwithstanding, out of the disaster at Pearl Harbor came recurring demands for a truly centralized permanent intelligence agency and increased participation of the civilian agencies in COMINT matters. Proposals for the establishment of a single United States intelligence agency became routine topics for discussion in the Joint Intelligence Committee of the JCS and in congressional investigations.

The combination of service competition, pending budget reductions, and high-level investigations foretold sweeping changes in the intelligence structure in the postwar years. The end of World War II signaled the beginning of the end of the exclusive military domination of the Army and Navy COMINT organizations. Civilian agencies now pressed for a much greater voice in the direction of U.S. COMINT activities.

Chapter II

The Military Services and the Joint Operating Plan, 1946-1949

Immediately following World War II, American policy makers looked for ways to achieve major reductions in the military budget. Despite the spectacular successes achieved by the Army and Navy COMINT organizations during the war, they quickly became prime candidates for reorganization and for major reductions in their resources. As the Pearl Harbor investigations continued, interest in intelligence matters also increased dramatically. For the first time, U.S. intelligence operations came under outside scrutiny. By 1946 service COMINT officials found that they were no longer able to act as free agents in making many of the basic decisions affecting their COMINT operations.

Their days of complete autonomy were numbered. Other fundamental intelligence relationships were also changing. Within the COMINT community, the addition of the State Department to the membership of the COMINT policy board changed not only the composition of the board but the scope of its intelligence interests as well. At the international level, the Great Britain-United States negotiations to extend COMINT collaboration into the postwar period were nearing completion. Finally, in the military itself, there now existed demands for closer cooperation between the Army and Navy COMINT organizations.

In addition, developments during World War II forced a new reassessment and push toward unification of the military services at the national level. Despite widespread agreement on the need for postwar organizational reform of the military services, there existed deep philosophical differences and suspicions among the services that could not be resolved easily. As debate progressed during this period it became clear that Congress would have to legislate a structure that would be acceptable to the military services.

All of these activities – foreign negotiations and unification – impacted on the COMINT structure that sought to achieve its own degree of unification within the intelligence organizations of the War and Navy Departments. As a principal means of achieving closer cooperation, the service COMINT organizations responded to these pressures by establishing a joint operating agreement. This new alliance called for a collocation of the Army and Navy COMINT processing activities in the United States, as well as cooperation in their COMINT collection and reporting programs. While the services remained organizationally independent, the joint operating agreement did call for a totally new managerial concept, namely, operating on the basis of shared" or "joint" control over a number of COMINT targets and resources. While this was a difficult period of adjustment for the COMINT services, they not only survived but made some significant COMINT contributions during this time.

As the services moved into the postwar period, they found that peacetime operations, rather than simplifying the conduct of their COMINT operations, brought new problems and highlighted even more the glaring disunity of the U.S. COMINT structure. By 1946 the harsh realities of the new situation began to hit home. Operationally, the services had lost their wartime targets of Germany and Japan, and the source of many spectacular successes. At the same time, the services were confronted with the specter of rapidly shrinking resources. Shortly after V-E and V-J Days, their parent headquarters ordered drastic reductions of their COMINT facilities. Demobilization actions were under way with dire consequences for the service COMINT operations.

While the services no longer had the urgency of a wartime situation to support their requests for resources, the likelihood of going through an extended period of austerity did have one practical effect. It forced the services to reevaluate their joint posture and to think more seriously about closer cooperation between their organizations. Because of the new public investigation of Pearl Harbor with its intensive probings into intelligence matters, the COMINT officials saw that they would be vulnerable, once again to charges of duplication of effort and inefficient use of resources if they continued to maintain totally separate and independent COMINT organizations.

Fortuitously, in the postwar period a new operational target emerged for the U.S. COMINT services. As the hostility of the Soviet Union toward the West became more apparent, the Army and Navy began to plan for a major adjustment of their COMINT coverage, to focus on Soviet targets. But the realignment was not all that simple. Some very fundamental questions existed concerning intercept and processing that could only be answered on a communal basis. For example, what were the new collection priorities? What were the new intelligence priorities? Who would establish these priorities? What were the interests and roles of the non-military consumers? How would the intercept and processing of the Soviet material be divided between the services? It became obvious that the service COMINT organizations, as constituted, could not answer these questions.

Because of these problems, the services perceived an immediate need for accomplishing some form of cooperation that went well beyond the scope of any previous efforts. The military authorities fully recognized that, at best, they had made only superficial progress toward the establishment of closer cooperation between their organizations in the production of COMINT. Earlier moves toward closer cooperation, dictated by wartime necessity, had been carefully designed to be limited in scope, as well as to avoid any interference with the primary interests of each service. The wartime agreements had accomplished little more than a basic division of labor and had avoided the real issue of establishing a centralized cooperative effort. In the main, the spirit of the earlier measures seemed to reflect an inherent attitude that cooperation in COMINT matters was a necessary evil, rather than any real conviction about the benefits of centralization or cooperation.

Seeking to shelter their vital COMINT functions from further budget reductions, the military authorities intensified their efforts to achieve closer cooperation and coordination between their COMINT organizations. The likelihood of further budget reductions and the question of survival of their separate COMINT organizations forced the two organizations together.

A few Army and Navy officials, aware of the magnitude of the COMINT successes during World War II, became the prime movers in the effort to preserve the Army and Navy COMINT structures. Three officers in particular stand out in the postwar era – Colonel W. Preston Corderman, USA, Colonel Carter W. Clarke, USA, and Captain Joseph N. Wenger, USN.[1] As early as 1943, these officers took the essential first steps in pressing for the preservation and fusion of the military COMINT resources. Over the next few years, they consistently took the lead in facilitating a dialogue between the services to foster the preservation of military COMINT resources. For the most part, they sought to promote service discussions covering a broad range of organizational relationships, such as division of responsibility on cryptanalytic tasks, the feasibility of joint operations, and possible ways to avoid unnecessary duplication. Each of these officers encountered varying degrees of opposition, sometimes from within their own service, and sometimes from the other service. Despite the continuing lack of enthusiasm encountered at various echelons of the military structures for consolidation, they had the foresight to view COMINT as a national asset that would be vital in meeting future U.S. intelligence needs. Corderman, Clarke, and Wenger never wavered in their single-minded determination to save the existing military COMINT structure from a dismantling process through budget cuts.[2]

Of the postwar intelligence machinery, the establishment of the Army-Navy Communications Intelligence Board was probably the most important component for the Army and Navy. With the creation of ANCIB in March 1945, Corderman, Clarke, and Wenger succeeded in establishing the nucleus for a structured, communal approach to the basic handling of COMINT matters – and in moving the services toward toward greater cooperation in their intelligence relationships. Operating with a very limited charter, ANCIB quickly emerged as a policy mechanism for the COMINT services and brought a new semblance of unity and order to the COMINT structure.

Reinforcing their goal of creating a self-governing mechanism for the COMINT agencies, the leaders brought about the establishment of an expanded policy board – the State-Army-Navy Communications Intelligence Board – in December 1945. The members established STANCIB as the primary governmental mechanism to coordinate and guide the activities of the COMINT structure and to assist in its reorganization during the postwar period.[3] In retrospect, the development of a strong role for the policy board stands as a tribute to the military leaders, particularly when recognizing that STANCIB was operating without an official charter.

Despite the fresh dialogue and new perspective on a broad range of COMINT matters, one critical element was still lacking within the COMINT structure that could prevent STANCIB from acting as the COMINT broker, at either the international or domestic level. While the services had achieved considerable progress in expand-

ing their dialogue at the policy level, they had not made similar progress in designing an operational plan that would enforce closer cooperation at the working level. Unless some additional leverage was brought to bear upon the services, the authorities recognized that the Army and Navy had gone about as far as they could – or would – go in achieving closer cooperation at the working level. Since voluntary merger was not likely to occur, direct intervention by higher authority was inevitable.

The proposal to merge the Army-Navy communication intelligence activities had been under periodic discussion by the services as early as 1942. The Army authorities generally supported the proposals for merger, while naval officers were unanimously opposed. For the Army, Major General George V. Strong, Assistant Chief of Staff, G-2, repeatedly expressed his strong support of the concept.[4] From the outset, however, the naval authorities opposed the concept of merger. Admiral Ernest J. King, Chief of Naval Operations, supported the position of the Navy's intelligence and COMINT officials that its COMINT operations should remain under exclusive naval control.[5]

The Navy's persistent opposition to the centralization of cryptologic resources stemmed, in large part, from its perception of its fundamental intelligence needs, as contrasted with those of the Army. The Navy considered that its intelligence requirements expressed statements of need for intelligence information of a strategic nature and of national-level interest, that could be properly handled only by a full-scale technical center under the operational control of the Navy. In contrast, the Navy perceived the Army's intelligence requirements as reflecting needs of a more limited nature, which were exploitable in the field at a tactical level. Leaving little room for negotiation on the issue, the Navy generally discouraged exploration of the concept of merger during the 1940s.

But the developments associated with the end of World War II brought about a general reopening of the feasibility of the merger concept. By V-J Day (14 August 1945), a number of new problems confronted the services that involved both operational and political considerations and that forced them to take a new look at their cryptologic organization.

At the same time, there existed a number of parallel developments at the national level that also seemed to threaten the COMINT services. Confronted with the reality of budget cuts, the services recognized that they would have to acquire new priority tasks in order to justify the continuance of their separate organizations. Moreover, the concept of centralization had acquired new credibility and momentum within the upper levels of the government. There existed growing pressures, emanating from both the presidential and congressional levels, to establish a new centralized intelligence agency and to accomplish, in some form, an integration of the military services. Once again, the issue for the military organizations related directly to the question of their continued existence.

Because of these factors, the Army and Navy command authorities moved to a position that clearly supported a merger of the COMINT services. A few days after the surrender of the Japanese, because of budgetary retrenchment actions and the loss of the major wartime targets, the Army and Navy command authorities clearly supported a merger of the COMINT services. An exchange of Army-Navy correspondence appeared to set the stage for accomplishing a merger action. General George C. Marshall, Chief of Staff, U.S. Army, in a letter of 18 August 1945 to Admiral Ernest J. King, Chief of Naval Operations, recommended a complete physical merger of the COMINT processing activities of the Army and Navy.[6] He proposed that the Joint Policy Board (ANCIB) study the proposals and develop specific recommendations on "how to insure complete integration." In his response of 21 August 1945, King expressed complete agreement with Marshall. King also noted that he had directed the Navy members of ANCIB to work with the Army representatives in the development of recommendations.[7]

With the Marshall and King exchange, the basic decision to merge the COMINT activities was made. All that remained was simply the matter of developing the ways and means for executing the decision for merger. On the surface it looked simple. Implementing the merger became the responsibility of ANCIB and its working committee, ANCICC.

On 28 August 1945 ANCICC responded by establishing a Subcommittee on Merger Planning (SMP). In its instructions to the SMP, ANCICC noted that the subcommittee had the task "of making recommendations in implementing the decision of General Marshall and Admiral King that the Army, Navy intercept, cryptographic, and cryptanalytical activities be merged under joint direction." The ultimate objective of the committee was to accomplish a prompt and complete merger of Army and Navy organizations in one location under ANCIB.[8]

One of the main tasks assigned to the committee was the selection of a site for the consolidated COMINT operations. Because of the need for direct exchange between producers and consumers, the committee concluded that the activity should remain in the Washington area. ANCICC presented an analysis and comparison of the Army site at Arlington Hall with the Naval Communications Annex (which it called the Mount Vernon Seminary). Because of its greater potential for expansion, the committee selected Arlington Hall as its first choice for the relocation of all COMINT activities. The Arlington Hall site of ninety-six acres was considerably larger than the Navy site of thirty-five acres. In its final report of 7 September 1945, however, the committee concluded that both sites should be retained, with

During the policy deliberations within ANCIB, Colonel Corderman, Chief, Army Security Agency, reiterated the traditional Army position for an immediate and complete physical merger of the two organizations. While Captain Wenger, head of OP-20-G, fully supported the concept of eventual consolidation, he personally espoused the view that merger should be accomplished as a gradual process in order to accommodate differences in organization and methods. These differences in approach, however, did not affect the final report that recommended a complete merger.[10]

But the situation soon changed within the Navy. The command authorities of the Navy, supporting the traditional naval view concerning central authority, overrruled Wenger at the eleventh hour. When ANCICC considered the final report on 12 September 1945, a new Navy submission completely nullified Wenger's earlier concurrence and indicated that even the concept of gradual consolidation went further than the Navy was willing to go. The Navy memorandum stated that

> *a full physical merger of Army and Navy communications intelligence activities does not seem desirable to the Navy. . .*

The memorandum also pointed out that

> *the Navy must retain complete control over all elements of naval command, so that the Navy will be free to meet its interests, solve its special problems . . . [and] must, therefore, have complete control over its operational intelligence.*[11]

The Navy's abrupt reversal of its earlier position brought to a complete standstill the entire move toward consolidation. On 26 September 1945 ANCICC closed out the activities of its Special Committee on Merger Planning and referred the matter to ANCIB for guidance.[12] ANCIB, however, had no authority to resolve the conflict between the services and looked instead to the departmental authorities for resolution. In trying to pick up the pieces, Marshall and King exchanged four additional letters during September and October 1945. But the letters reflected no change of positions, as each simply reiterated the previous position of its intelligence service, with no specific suggestions offered for compromise. [13]

On 14 October 1945 King reported to James V. Forrestal, secretary of the navy, that he and General Marshall continued to agree that the coordination of signal intelligence activities could be improved, but they had not achieved a solution satisfactory to both services. King noted that both services agreed that the processing of some types of traffic should be jointly undertaken, but the exact manner in which this might be accomplished remained unresolved.

> *The Army favors a complete merger of our cryptanalytic Units under one director, whereas the Navy, desirous of insuring its control of operational intelligence essential to naval commands, does not favor a complete merger but would rather effectuate the desired results by joint effort under joint direction.*[14]

By December 1945, however, new participants appeared on the scene. General of the Army Dwight D. Eisenhower replaced General Marshall as Chief of Staff of the Army, and Fleet Admiral Chester W. Nimitz relieved Admiral King as Chief of Naval Operations. In a positive move toward solution of the problem, Eisenhower reopened the issue in a letter of 2 January 1946, suggesting that "we should make a fresh start on this entire subject." Remarking about their earlier experiences as commanders of combined forces, Eisenhower commented that "we both know how vital it is to resolve any differences of opinion and to achieve complete integration as soon as possible." His proposal was very simple. He proposed that the Army and Navy members of ANCIB should either solve the problem by themselves or develop alternative proposals for decision by Eisenhower and Nimitz.[15]

Nimitz readily accepted Eisenhower's suggestion for making a fresh start on the issue of how to integrate and coordinate the COMINT activities of the Army and Navy. As evidence of a softening of the Navy's position, Nimitz instructed the Navy members of ANCIB to consider the problem with open minds, free of any restrictions stemming from earlier policy guidance.[16]

With the new push from Eisenhower and Nimitz, the COMINT officials of the Army and Navy began to reassess their earlier positions. In lieu of having a solution directed by higher authority, both services obviously preferred to solve the problem at the COMINT level. Even the monolithic Navy, after having derailed the earlier efforts toward merger, indicated a surprising new willingness to go along with the move toward consolidation. As the spokesman for the Navy COMINT organization, Wenger pressed again for the concept of gradual consolidation as representing an attainable solution.[17] Similarly, the Army advocates of consolidation ultimately modified their earlier position on merger and came to acknowledge that the objective of a complete merger would have to be deferred for a later date. Corderman, one of the main proponents of merger, had also insisted previously that the senior joint official selected to head the merger should be identified as the "Director" rather than "Coordinator." Before the new negotiations were over, however, Corderman would yield on this point as well.[18]

The Navy resubmitted Wenger's earlier concept paper as its new bargaining position. In effect, the naval authorities supported a position that went one step further in the move toward consolidation but that still fell short of complete merger. The Navy officials would support a concept described as the establishment of an "effective working partnership" between the Army and Navy. In a modification of Wenger's earlier paper, the Navy proposed the establishment of a new position, the Coordinator of Joint Operations (CJO). The CJO would not function as a czar with unlimited authority, but rather would have the responsibility for facilitating interservice coordination and cooperation. Under the terms of the new Navy proposal, the services would function as coordinating but independent organizations. Some joint operations would be established. Further, the services would ensure a continuous cooperation an exchange of information on all other COMINT problems. Policy control of the structure would be vested in a Joint Policy Board (ANCIB-STANCIB) that in turn would reflect the interdepartmental authority of the chief of staff, U.S. Army, and the commander in chief, U.S. Fleet and Chief of Naval Operations.[19]

By early 1946, the British-United States of America Agreement (BRUSA) negotiations, initiated in 1945 to establish postwar collaboration in COMINT between the two nations, were nearing completion. Since the concept of BRUSA collaboration was predicated in part on the existence of centralized controls of COMINT activities within both countries, the approaching ratification and implementation of the agreement brought a new, compelling urgency for the United States to put its own house in order. These international considerations, coupled with the departmental pressures stemming from the Eisenhower-Nimitz exchange, prompted new discussions in STANCIB concerning possible ways to merge the Army and the Navy COMINT organizations.

On 13 February 1946 STANCICC considered at length the earlier Navy proposal for closer cooperation of the Army-Navy communications intelligence activities.[20] Moving very quickly on the issue, on 15 February 1946 STANCIB approved in principle the framework for a new concept of Army-Navy cooperation in COMINT.

The Navy's insistence on establishing a "Joint Effort under Joint Direction" prevailed in the discussions of the COMINT policy board. STANCIB accepted the framework for a new period of Army-Navy cooperation in COMINT, based on the Navy's earlier proposal of "joint" but "separate" COMINT activities.[21]

The STANCIB decision ruled out the possibility of any actual merger of Army-Navy COMINT processing activities. Instead, the services would now undertake new initiatives to achieve closer cooperation on all phases of the COMINT process. This improved cooperation would be achieved by establishing closer working liaison day-to-day in the functional areas of intercept, analysis, and reporting. Integration of technical personnel from the opposite service would also take place – primarily on analytic problems – at Arlington Hall and the Naval Communications Station. The new agreement, however, pertained only to the collection and production of information from foreign communications. It excluded such intelligence functions as estimates or the dissemination of COMINT information as finished intelligence.

The COMINT organizations would coordinate their activities but would remain totally independent organizations. In addition to the integration of Army-Navy personnel on certain analytic problems, STANCIB divided the Army-Navy responsibility for some targets along the traditional lines and identified others as a "joint" responsibility, to be placed under the direction of the new Coordinator of Joint Operations. To implement this new concept of Army and Navy cooperation, STANCIB directed the chiefs of ASA (Army Security Agency) and OP-20-G to draw up the details of a plan and statements of general principles governing the roles and responsibilities of the services and the Coordinator of Joint Operations.[22]

By approving this new concept of "partnership," STANCIB succeeded in keeping its

efforts to reorganize the U.S. COMINT structure in tandem with the progress of the BRUSA negotiations. By 1946 STANCIB, although lacking a national charter, had succeeded in positioning itself as the primary U.S. authority and spokesman for policy negotiations with foreign nations on COMINT matters. At the same time, STANCIB also greatly enhanced its stature as the central organization for promoting closer cooperation between the U.S. services.

Lieutenant General Hoyt S. Vandenberg, STANCIB chairman

On 5 March 1946 the U.S.-U.K. representatives formally signed the British-United States of America Agreement, which authorized continued postwar collaboration in COMINT matters on a governmental basis. Lieutenant General Hoyt S. Vandenberg, STANCIB chairman, signed the agreement for the United States, and Colonel Patrick Marr-Johnson, representing the London Signals Intelligence Board, signed for the United Kingdom.[23]

As a follow-up to the BRUSA Agreement, a "Technical Conference" took place in London several months later. The primary task of this conference was to develop the overall blueprint for the development of technical appendices to the agreement. Over the next few years, this initial effort resulted in the development of a number of appendices to the BRUSA Agreement, which governed such areas as security, collection, liaison, and other aspects of collaboration.[24]

On 22 April 1946, six weeks after ratification of the BRUSA Agreement, STANCIB issued the

"Joint Operating Plan" (JOP). The JOP also became known as the "Corderman-Wenger Agreement," named for the principal Army and Navy negotiators (Colonel W. Preston Corderman, USA, and Captain Joseph N. Wenger, USN).[25]

Colonel W. Preston Corderman, USA

Captain Joseph N. Wenger, USN

As an integral part of the plan, STANCIB approved an expansion of its own charter. This change provided for the establishment of a fundamentally new position, the Coordinator of Joint Operations. The new coordinator, it was hoped, would become the driving force in unifying the COMINT structure. According to the charter, the CJO would function in a dual capacity and under dual command lines. First, the CJO would function as an executive for STANCIB, and thus would be responsible for directing the implementation of STANCIB's policies and directives relating to intercept and processing tasks, as well as

for all joint projects with other U.S. and foreign intelligence agencies. In addition to his STANCIB role, the CJO would acquire a new leadership role within the Army and Navy COMINT structures on day-to-day operations involving joint tasks. Organizationally, the CJO would have dual subordination lines, reporting to STANCIB as the CJO, and to his individual service in his capacity as chief of a military COMINT organization. [26]

Under the Joint Operating Plan, there were two key positions that governed the conduct of COMINT operations. These were the CJO and the chairman of the working committee (STANCICC) of the COMINT Policy Board (STANCIB). The chief of the Army and Navy COMINT organizations rotated yearly as the incumbent of each position. This rotation of the senior service officials gave each service a continuing and powerful voice in the "coordination" and "policy" roles.[27]

The STANCIB-STANCICC structure served to facilitate resolution of some disagreements, but there were still problems. The rule of unanimity still prevailed on the policy board as well as on its working level committee. Thus, whenever STANCIB-STANCICC failed to reach a unanimous decision on an issue, it remained unresolved.

Vandenberg, as chairman of STANCIB, recommended that the first coordinator be selected from the Army because of its wartime COMINT activity.[28] Following this recommendation, STANCIB selected Colonel Harold G. Hayes, chief of the Army Security Agency, as the first CJO on 1 May 1946. The operating chiefs of ASA and OP-20-G came responsible to Hayes for accomplishing those tasks that he allocated to them.[29] Hayes was "to coordinate," however, and not "to direct." It was an important distinction.

Under the plan, the Army and the Navy maintained their independent COMINT organizations.

Colonel Harold G. Hayes, Chief of Army Security Agency, first coordinator of Joint Operations.

The Joint Operating Plan directed that the responsibility for each COMINT problem be allotted to the Army or Navy in such a way as to prevent any duplication or overlapping of effort. Thus each service continued to control a large percentage of its own intercept and processing capacities. Each service also performed "tasks of common interest," such as work on weather targets. Although the CJO allocated these tasks to the services, the actual intercept facilities remained under the tasking and control of the services. The CJO, however, did control and coordinate the intercept coverage and reporting on the "Joint Tasks."

The term "joint" applied generally to all tasks not strictly Army or Navy. These tasks represented areas of special interest. The CJO exercised his authority over these tasks by establishing a committee'on group structure, designed along functional lines, that reported to him. These areas included intercept, processing, and liaison activities.[30]

Administratively, three subordinate groups assisted the CJO: a Joint Intercept Control Group (JICG), a Joint Processing Allocation Group (JPAG), and a Joint Liaison Group (JLG).[31] A deputy coordinator served as the chief of each group. While the CJO was to use existing facilities whenever feasible, each service also assigned personnel to him for his own staff support. This included clerical, administrative, and analytical assistance. The coordinator's senior assistant was from the opposite service and normally served as chief of the JPAG. Captain Charles A. Ford, USN, served as the first chief of JPAG. The officer in charge of the Joint Liaison Group was also from the opposite service. Commander Rufus L. Taylor served as the first chief of the JLG under Hayes. The officer in charge of the Joint Intercept Control Group was from the same service as the coordinator; Lieutenant Colonel Morton A. Rubin, USA, served as first chief of the JICG.[32]

The mission of the Joint Intercept Control Group was to develop a plan for intercept coverage that would provide intelligence of maximum value to the consumers. The JPAG allocated processing responsibilities to the Army and Navy. As the U.S. overseer of foreign liaison, the JLG arranged for and supervised U.S. working arrangements in COMINT with the United Kingdom and Canada. In addition, six standing subcommittees of the COMINT policy board served as advisory committees in the areas of intercept, direction finding, cryptanalytic research and development, communications intelligence and security, traffic analysis, and collateral information. In this complex structure of functional groups and STANCICC subcommittees, the deputy coordinators of the groups and chairmen of the STANCICC submmittees were under the direct supervision of the CJO.[33]

After the establishment of the JOP in April 1946, additional organizational changes took place affecting the STANCIB structure. After examining a draft of the BRUSA Agreement, J. Edgar Hoover, director of the FBI, expressed an interest in obtaining membership on STANCIB.[34] Adding the FBI to its membership on 13 June 1946, the board and its subordinate committee became the United States Communication Intelligence Board (USCIB) and the United States

Communication Intelligence Coordinating Committee (USCICC).[35] When Lieutenant General Hoyt S. Vandenberg, assistant chief of staff, G-2, became the second Director of Central Intelligence in June 1946, USCIB agreed to expand its membership once again by including the DCI as the representative of the newly established Central Intelligence Group (CIG).[36] (The CIG came into existence on 22 January 1946.)

As the membership of the policy board increased, the civil agencies such as the Department of State, FBI, and the Central Intelligence Agency (CIA) began to participate in the activities of USCIB and the JOP of the Army and Navy. (The National Security Act of 1947 established the CIA on 18 September 1947, superseding the CIG.) As members of USCIB, however, they participated only as observers in the activities of the Joint Intercept Control Groups and the Group. From 1946 to 1949, these committees of USCIB and the CJO were the primary mechanisms available to the intelligence consumers for expression of their intelligence priorities and specific requirements for COMINT information.[37]

A major problem area for the JOP proved to be intelligence requirements. The military services continued to handle their requirements basically on a service-to-service basis. For example, the Army G-2 tasked the Signal Security Agency for its COMINT requirements, with the same parallel applying to the Navy. However, the area of "joint" interests remained poorly defined, both for military targets and for other broad targets of interest to civilian agencies. Despite the organizational change in the COMINT structure, the civilian agencies quickly recognized that they still had no real voice or representation in the adjudication or establishment of intelligence priorities.[38] Changes were taking place, however, that would give a new prominence to the consumer role, as well as a greater participatory role for the civilian agencies in the operations of the COMINT structure.

After operating for three years under a purely interdepartmental charter, USCIB acquired a new national charter in 1948. The new National Security Council Intelligence Directive Number 9, "Communications Intelligence," established USCIB as a national COMINT board reporting directly to the National Security Council rather than to the military departments. The charter, however, was not appreciably strengthened, and still reflected a preponderance of military membership. But the change of subordination, coupled with the establishment of CIA in 1947, meant that the military COMINT community could no longer act in a totally independent manner.[39]

Under the JOP, the primary vehicle for the dissemination of COMINT to consumers was the published translation or bulletin, issued in a standard format prescribed by the JPAG.[40] The Army and the Navy generally issued separate bulletins on their respective targets. Bulletins on joint-interest targets were published as joint Army-Navy products. Within this overall framework also existed a number of separate reporting series for major categories of information such as Soviet COMINT. As provided in the BRUSA Agreement, bulletins were exchanged with GCHQ.[41]

The creation of the JOP marked the introduction of major changes involving producer and consumer relationships. These changes provided the consumers with greatly expanded technical information in COMINT reporting and granted them greater access to COMINT activities. At its thirtieth meeting, 27 April 1948, USCIB approved a CIA request for greater access to COMINT activities. [42] This decision authorized all of the consumer agencies to receive unfinished products considered necessary for the fulfillment of their mission of producing finished intelligence. In addition, consumers now had the option of placing indoctrinated representatives within the COMINT production organizations of the Army and the Navy. The ground rules governing these relationships required that specific arrangements be worked out in each case, primarily through working-level contacts or through the service COMINT authorities. Lacking resolution via these channels, the consumer still had the option of referring the matter to USCIB for further consideration.

During this period, any evaluative process or further dissemination of COMINT became the responsibility of each consumer. Generally, the agencies accomplished this by collating the COMINT with other intelligence information and by preparing special fusion reports containing both COMINT and other intelligence sources, Since most of the USCIB members prepared their own community-wide reports, this resulted in a wide variety of publications. These included a daily summary published by the Department of State; the Army's summary; and the Navy's "Soviet Intelligence Summary," both issued weekly; and various other special reports issued by the Army, Navy, and CIA.[43]

To assist the agencies in their evaluation of COMINT, the Army-Navy COMINT bulletins included specific data related to the origins of intercept. In addition, the consumers, on request, could receive unfinished COMINT products considered necessary for their own evaluation.[44] Each consumer also prepared its own estimates. This often resulted in a number of different intelligence estimates on any one subject – with no organization producing a consolidated estimate. Thus, the difficulties associated with centralization of the COMINT organizations extended to the entire intelligence process, and to the consumer membership of USCIB such as CIA, State, Army, Navy, and Air Force.

In summary, in response to growing national pressures, and as a principal means of achieving closer interservice cooperation, the COMINT services established a joint operating agreement rather than undertake a merger of their separate COMINT organizations. The mechanics of this new alliance called for a collocation of the Army and Navy COMINT processing activities in the

United States, as well as for major organizational changes in their collection and reporting tasks. The move to establish joint service operations reflected a realization of their increasing interdependence as well as of the inevitability of still further changes in the management of COMINT resources. While the services remained organizationally independent, the joint operating agreement called for the introduction of a totally new managerial concept for the services, namely, operating on the basis of "shared" control over COMINT resources.

Because of the magnitude of the governmental changes from 1946 to 1949, the JOP represented a period of great adjustment for the COMINT services, as well as for the entire intelligence community. At a time when the services and Congress were still debating the unification issue, the creation of the JOP occurred harmoniously and by mutual agreement of the Army and Navy. By collocating and integrating their COMINT processing centers at Arlington Hall and the Naval Security Station, the JOP achieved a level of interservice cooperation never previously accomplished by any military organization.

The concept of the JOP was very simple. It called for the services to act as coordinated but independent agencies. In developing the blueprint for the JOP, the primary objective was to devise a structure that by its very nature would promote greater cooperation and dialogue between the military COMINT organizations. At the same time, Army and Navy authorities insisted on maintaining their separate identities and organizations. In the implementation, some elements of the reform process proved to be highly effective as the services actually began to cooperate on COMINT matters. The plan was seriously flawed, however, as it stemmed from a proliferation of military command lines and created a rash of new bureaucratic channels requiring coordination. For example, the CJO functioned under dual subordination, reporting to USCIB as the CJO, and to his individual service as the chief of a military COMINT organization. This dichotomy of authority proved to be not only conceptually unsound but detrimental to the timeliness of COMINT operations.

Another organizational drawback of the JOP structure was that it called for a large committee structure to work on functional matters of an operational nature and to operate under the aegis of the CJO. While these committees soon became pivotal coordination points, they also became representative of a "management by committee" syndrome with all the traditional weaknesses of a committee process, such as procedural delays and the inability to make timely decisions.

In a more positive vein, however, the JOP merits high marks for some very significant accomplishments. Considering the innate service opposition to the concept of merger of their COMINT processing activities, it was a major achievement that the services agreed to adopt a concept of collocation and integration of any kind. By establishing a form of quasi-consolidation, and operating on the principle of gradual change, the JOP constituted a compromise. It provided a logical transitional structure for the services as they entered the postwar period. In addition, the JOP accomplished sweeping organizational changes for the services, such as realignments of operational elements, personnel, and mission without causing a catastrophic upheaval of their operational missions and functions.

By achieving a nominal degree of centralization of U.S. COMINT efforts, the JOP facilitated the ratification of the BRUSA Agreement. Without this tightening of controls over the Army and Navy COMINT activities, the BRUSA Agreement would not have been attainable, as it implicitly called for greater centralized control of COMINT activities within both nations.

Operationally, the JOP facilitated the realignment of U.S. COMINT targets for peacetime, including the assumption of a new national target, coverage of the Soviet Union.

Finally, by virtue of the cosmetic fusion of the two services via the JOP, the Army and Navy COMINT organizations were largely able to survive the chaotic period of demobilization and budget reductions following the war. Despite heavy attrition, these two organizations maintained a solid operational base, along with a cadre of professional talent, for the next cycle of reorganization.

Chapter III

The Emerging National Intelligence Structure and the United States Communications Intelligence Board, 1946-1949

The period 1946-1949 marked the beginning of efforts at both the presidential and congressional levels to view intelligence matters as a national responsibility. As a first step toward the centralization of U.S. intelligence activities, in 1946 President Truman established a National Intelligence Authority (NIA) and a Central Intelligence Group (CIG). One year later, in the summer of 1947, Congress passed the National Security Act, which resulted in a further realignment of the national intelligence structure. This landmark legislation disestablished the NIA and CIG and created a National Security Council and a Central Intelligence Agency. The act also provided for a secretary of defense and a realignment of the U.S. defense organization. These changes were only the beginning.[1]

The COMINT structure felt the impact of these changes almost immediately. By 1948 the United States Communications Intelligence Board had a new official charter that made it subordinate to the National Security Council. This was the result of a series of controversies over the jurisdiction of COMINT issues. The participants in making these changes were James V. Forrestal, secretary of defense; Admiral Roscoe H. Hillenkoetter, DCI; Admiral Sidney W. Souers, executive secretary, NSC; and the members of USCIB, Army, Navy, State Department, and the FBI. The main battle was fought over what organization should "control" USCIB, as well as the various components of the COMINT structure. This was essentially decided when the National Security Council issued USCIB's new charter in 1948.

As an agency of the NSC, USCIB now acquired a greatly enhanced policy role in the intelligence community. In addition, the new charter recognized that the civilian agencies had a vital part to play in the development of national intelligence policy and in the establishment of national intelligence priorities. A review of the debates during the reform process strikingly illustrates the major changes in the U.S. intelligence structure and USCIB. It also shows the sharp divisions that existed among the members of the intelligence agencies.

At the end of the Second World War, there existed no semblance of unified control over the conduct of U.S. COMINT activities – nor was there any external body that had sufficient authority to provide guidance and direction to the extremely powerful military COMINT structure. The primary management controls over the COMINT functions came from three sources, namely, USCIB, and the headquarters of the Army and Navy COMINT organizations. A creation of the military departments, USCIB served as the nominal policy authority for COMINT matters, while the military departmental authorities provided a number of internal controls relating primarily to administrative and budgetary matters.

By 1946 USCIB's membership included representation from the Army, the Navy, the FBI, the Department of State, and the CIG. Because of the dual representation accorded each service, the Army and Navy played dominant roles in the activities of USCIB and its many subcommittees. For example, the assistant chief of staff, G-2, and the commanding officer, Army Security Agency, represented the Army. The assistant chief of staff for Intelligence, U.S. Fleet, and the director of Naval Communications provided representation for the Navy. Each had a vote; the other members of USCIB – FBI, State, and the CIG - had only one representative and one vote each.[2]

James V. Forrestal, first secretary of defense, 1947-1949

*Admiral Roscoe H.
Hillenkoetter
director of Central
Intelligence*

*Admiral Sidney W. Souers
executive secretary, NSC*

During this early period, the Army provided coverage of the intelligence interests of the Army Air Force. In May 1947, however, the Army Air Force obtained its own separate representation. On 7 May 1947 USCIB invited the commanding general, Army Air Force, to appoint representatives to USCIB and to its subordinate committee, the U.S. Communications Intelligence Coordinating Committee (USCICC). On 29 May 1947 the Army Air Force designated Major General George C. McDonald, assistant chief of staff, Army Air Force, as its representative to USCIB. Brigadier General Francis L. Ankenbrandt, communications officer, Army Air Force, became the representative to USCICC.[3]

USCIB's early charter of 31 July 1946 stated that USCIB would meet only to decide questions of major policy or to consider matters that its working committee, USCICC, could not resolve.[4] Procedurally, USCIB elected its own chairman, usually the senior member, who served for one year. It met at the will of the chairman, or subject to the concurrence of a majority, and the request of any member. The rule of unanimity governed both USCIB and USCICC discussions. When agreement could not be reached, the only option was to refer the matter to higher authority within the members' departmental organizations. In short, USCIB functioned solely in the capacity of providing guidance and coordination for the services, which they were free to accept or reject.[5]

Despite the limited nature of its founding document, the Marshall and King Agreement of 1945, USCIB achieved considerable progress as the self-appointed authority for COMINT matters. By expanding its membership to include the FBI, State Department, and the CIG, USCIB had become a joint military and civilian board, with increasing involvement in the activities of the entire U.S. COMINT structure. Unfortunately, the governing documentation of USCIB had not kept pace with the scope of its activities.

The documentation relating to membership changes was reflected only in the interdepartmental correspondence and in the subsequent updates of the USCIB Organizational Bulletin. The actual enabling document remained unchanged with no attempts made to amend or to reissue the document at a higher level to reflect the broader role and responsibilities of USCIB.[6]

Within the government, the nature of the intelligence process itself tended to foster the continuing independence, if not isolation, of the military COMINT activities. Long before Japan's attack on Pearl Harbor, the services traditionally handled communications intelligence, as well as all matters related to the COMINT process, as extremely sensitive information, releasable only by a strict interpretation of the "need-to-know" principle. The advent of World War II reinforced this well-established practice as the dual requirement for secrecy and anonymity of organizations intensified. Even within the COMINT structure itself, ULTRA and MAGIC materials, for example, were always strictly guarded and controlled, with only a very limited number of people being aware of these sources and their origins.[7] Distribution was made only to an authorized list of recipients, which included key military commanders and a few top officials of the armed forces, the FBI, and the Department of State.

These dual factors – the rigorous compartmentation practices and the absence of any dominant central authority – tended to foster an atmosphere of independence and isolation among the services. Except for administrative and budgetary guidance provided by their own departmental authorities, the Army and Navy COMINT organizations generally remained sheltered from critical review by external authority. They continued to be free agents and made their own decisions concerning their intelligence priorities and intercept coverage. When the USCIB became more active in these areas, it soon found itself powerless to direct the COMINT activities, primarily because of the inherent weaknesses of the USCIB charter and the military domination of the structure. As a result, the services generally encouraged and facilitated continuation of the status quo, thereby assuring themselves of almost complete freedom of action in the running of the COMINT business.

The same amorphous situation applied to general intelligence collection as well. No single organization had the overall authority and responsibility for the oversight of matters relating to the collection, analysis, and dissemination of all-source intelligence information; nor did any central organization exist with authority over the many and diverse producers of intelligence information. Each military department had its own intelligence branch and producing elements, as did the Departments of Justice, Treasury, and State.[8]

Long before V-E and V-J Days, considerable discussion and debate took place within the Joint Chiefs of Staff and the military departments, as well as in State and the FBI, concerning the organization and nature of the U.S. intelligence gathering apparatus for the postwar period.[9] By the fall of 1945, President Truman, known to favor a concept of centralized control, had already received several proposals for the establishment of a peacetime intelligence structure. He had drafts from the War Department, the Joint Chiefs of Staff, the Office of Strategic Services, the two military services, the Bureau of the Budget, and the Department of State. A number of these plans, primarily from the military organizations, recommended the establishment of a single centralized agency, but differed considerably on the designation of the controlling authorities. Proposals from the Department of State and the Bureau of the Budget generally recommended a status quo approach that would permit the intelligence offices of all departments to remain independent agencies, with no centralized agency to be established. Under this concept, however, there would exist a number of

advisory committees to assist a National Intelligence Authority in providing guidance to the intelligence activities of the various departments.[10]

Within a few months, Truman acted to centralize the intelligence structures. In a letter dated 22 January 1946 to the secretaries of state (James F. Byrnes), war (Robert P. Patterson), and navy (James V. Forrestal), Truman established a National Intelligence Authority (NIA) and ordered the secretaries to establish a Central Intelligence Group.[11] One day later, Truman appointed Rear Admiral Sidney W. Souers, the deputy chief of Naval Intelligence, as the director of the CIG, and the first Director of Central Intelligence (DCI).

The membership of the NIA consisted of the secretaries of state, war, and navy, and the president's personal representative, Rear Admiral William D. Leahy, his chief of staff.[12] Based on the presidential directive, its mission was to ensure "that all Federal foreign intelligence activities are planned, developed and coordinated so as to assure the most effective accomplishment of the intelligence mission related to the national security." Truman further directed the secretaries "to assign persons and facilities from your respective organizations, which persons shall collectively form a Central Intelligence Group and shall, under the direction of a Director of Central Intelligence, assist the National Intelligence Authority."[13]

The initial CIG was a unique structure. It had no assets or resources of its own. As a "collective interdepartmental group," it operated within the limits of the resources provided by the State, War, and Navy Departments. As an interdepartmental coordinating group, the CIG was responsible for planning and coordinating the government's intelligence activities and for evaluating and disseminating intelligence. Because of its limited charter and its limited resources, the CIG proved to be an interim, ineffective structure, However, by the mere establishment of the CIG, Truman had succeeded in setting in place the initial framework for the development of a centralized intelligence structure.

In addition, Truman's letter directed the establishment of the first postwar Intelligence Advisory Board (IAB), which served in an advisory capacity to the DCI. The IAB's membership consisted of representatives from the principal military and civilian agencies of the government that had functions related to national security, as determined by the National Intelligence Authority. The initial membership from the four departmental intelligence services consisted of Colonel Alfred McCormack, USA, State; Lieutenant General Hoyt Vandenberg, Army; Rear Admiral Thomas B. Inglis, Navy; and Brigadier General George C. McDonald, Air Force. [14]

Eighteen months later, still further changes took place at the national level, all of which greatly strengthened the initial presidential efforts toward centralization. Following months of intense debate by congressional and departmental authorities over the nature of America's security and defense force, Congress passed and President Truman signed into law the National Security Act of 1947 on 26 July 1947. [15]

With the passage of the National Security Act, also known as the Unification Bill, Congress abolished the National Intelligence Authority and its operating component, the Central Intelligence Group. In their place, the National Security Act established the National Security Council and the Central Intelligence Agency. Among other things, the Act created a National Military Establishment (NME) and three coequal departments of the Army, Navy, and Air Force within the Defense Department. It also established the United States Air Force, a War Council, and a Research and Development Board. [16]

The mission of the newly established National Security Council was to serve in an advisory capacity to the president in matters concerning the integration of domestic, foreign, and military policy. Its permanent membership consisted of the president, vice president, the secretary of state (Dean G. Acheson), secretary of defense (James V. Forrestal), and the chairman, National Resources Board. Optional attendees, depending upon the subject matter, were the secretaries and under secretaries of other executive departments and the military departments, the chairman of the Munitions Board, and the chairman, Research and Development Board. [17] The initial membership of the NSC did not include the DCI.

The effect of the National Security Act of 1947 on the direction and organization of COMINT activities was not at first discernible. Under the act, the National Security Council (NSC) received the broad mission of advising the president on matters of policy concerning national security. The new CIA, headed by the DCI, acquired a statutory base and became an independent agency under the NSC. This new intelligence agency had the stated responsibility for correlating, evaluating, and disseminating national intelligence; for rendering intelligence services to other agencies; and for advising the NSC in intelligence matters. [18] Under this vague charge, however, the relationship of the CIA to the COMINT-producing agencies remained obscure. It would be many years before CIA would assume its full responsibilities as initially conceived in the National Security Act.

From a military perspective, however, the National Security Act of 1947 had an immediate impact on the COMINT community. After the establishment of a separate Department of the Air Force, the Army Security Agency (ASA) continued to provide COMINT support and other support services to the Air Force on a transitional and interim basis. On 3 June 1948 ASA established an Air Force Security Group (AFSG) as a unit within its own Plans and Operations Staff. The AFSG,

President Truman signing amendments to the National Security Act of 1947

composed exclusively of Air Force personnel, operated as an Army element on an interim basis, with the mission of assisting the Air Force in its gradual assumption of COMINT responsibilities. By January 1949, following the approval by the secretary of defense of an Army and Air Force Agreement, the Army transferred personnel, facilities, and missions to the new Air Force Security Service (AFSS). On 1 February 1949 Colonel (later Major General) Roy H. Lynn, USAF, became the first commander of the USAFSS.[19]

Major General Roy H. Lynn, USAF

USCIB also changed and became much more active in COMINT matters. Two months after the passage of the National Security Act, USCIB began to hold regular monthly meetings. This action came about mainly at the urging of Admiral Earl Stone, one of the Navy members of USCIB, at the twenty-first meeting of USCIB on 4 November 1947. At the same meeting, Admiral Roscoe H. Hillenkoetter, who became the third DCI on 1 May 1947, raised the issue of USCIB's outdated charter. He commented that the existing interdepartmental charter in the form of an agreement by Marshall and King did not include the more recent members of USCIB – the CIA and the Department of State. Hillenkoetter, obviously eager to use his new departmental authority, pressed for the issuance of a higher-level charter to be issued in the form of an executive order by the president. Following Hillenkoetter's urging, USCIB directed its subordinate committee, USCICC, to examine the question of the proposed charter and to prepare several alternate versions.[20]

No disagreement existed among the military services on the need for a new charter. They recognized that USCIB had been performing as a national-level structure, but without the benefit of a charter commensurate with the scope of its activities and responsibilities. In the area of foreign collaboration in particular, they considered that USCIB needed a more authoritative charter because of the growing U.S. involvement in foreign COMINT relationships. When General Vandenberg signed the agreement for USCIB in 1946, he did so only after receiving prior approval of the agreement from Admiral Leahy, chief of staff to President Truman.[21] Consequently, the Army, Navy, and Air Force representatives enthusiastically supported the view that USCIB's authority should stem from a national-level issuance rather than from the existing military documentation. Despite the apparent agreement, the task would not be accomplished easily. Many jurisdictional and political problems now surfaced, the foremost of which related to the basic question of control."[22]

During this period, a membership change also took place in USCIB. In November 1947 the FBI voluntarily withdrew from the USCIB. In his letter of withdrawal, J. Edgar Hoover noted that "USCIB's discussions have been primarily concerned with methods of policy formulation within the Armed Services." At the twenty-first meeting of USCIB, however, the remaining members expressed the view that as a practical matter, the FBI was withdrawing from the cryptanalytic field primarily because of a lack of funds.[23]

Since the FBI had never been a very active participant in the activities of USCIB, its withdrawal had no immediate impact on the intensity of the charter discussions. The representatives of the State Department and the CIA continued to play the major role in representing the "civilian" interests in the USCIB deliberations. The representatives of State, in particular, were consistently articulate and persuasive in the presentation of their positions. Their position,

however, was not always predictable, as they frequently joined the military members in opposition to CIA efforts to acquire greater control of all intelligence operations.[24]

The competing interests of the board's military and civilian membership occupied center stage in the USCIB deliberations. Each group pressed for the supremacy of its own interests in the realignment of the COMINT policy board and in the competition for scarce COMINT resources. The battle lines reflected the JCS interests as opposed to the interests of the State Department and the new CIA.[25] The controversy over the proposed new USCIB charter continued for the next seven months. The basic issues were (1) what organization should be the parent body of USCIB and (2) what should be the role of USCIB – should it control or coordinate the national communications intelligence effort?

On the first issue, the parent body of USCIB, the Navy, the State Department, and the DCI, preferred that the NSC or the secretary of defense have overall supervision. The Army and the Air Force favored the JCS as the ultimate authority. A compromise, originally suggested by the Navy, and later proposed formally by the State Department, was the creation of a "Committee of Four" as an appellate body for USCIB. Such a committee would be composed of the three service chiefs of staff and the under secretary of state. USCIB accepted this proposal on 3 February 1948.[26]

On the second issue, whether USCIB should control or coordinate intelligence activities, there was a wide divergence of views among the members of USCIB. At the USCIB meeting of 6 January 1948, the members analyzed at length the nature of USCIB's political power over COMINT. Navy officials supported the view of the DCI and State that USCIB should exercise control authority over COMINT because of the national aspects of the COMINT effort. They uniformly identified the COMINT function as representing an intelligence resource of national potential rather than one of purely military interest. They further believed that the services had failed in their efforts to improve interservice coordination, and they strongly favored granting USCIB control of facilities assigned to common interest targets. Hillenkoetter even insisted that "COMINT agencies are military units in a limited sense because the civilian departments and agencies have an equal interest in the COMINT product."[27] Hence, it was a national resource and should be controlled by USCIB.

While the Navy strongly opposed the centralization of COMINT resources, it moved in the other direction in terms of how the COMINT policy board should be subordinated. For example, during the preliminary U.S. actions associated with the BRUSA Agreement of 1945, the Navy expressed concern in STANCIB meetings about the limited scope and questionable legality of the STANCIB charter, particularly in the area of foreign relations. Similarly, in recognition of the unique intelligence requirements of the CIA and the Department of State, the Navy had cited the need for a neutral, national forum to allocate intelligence priorities. In effect, the Navy had consistently argued for a stronger USCIB to reflect national interests. Considering the centralization of intelligence resources and the subordination of USCIB to be two separate issues, the Navy never considered its position on these issues to be contradictory.

The Army and Air Force members of USCIB, however, were equally adamant in their opposition to any arrangement that would give USCIB primary control over COMINT functions. They objected to placing USCIB, with its civilian members, in a chain of command between the military authorities and their operating COMINT agencies. Nevertheless, the Army and Air Force supported a lesser coordinating role for USCIB, with the primary control over the COMINT activities remaining in each department and agency. In fact, the Army officials preferred to dilute the

authority of USCIB even further. They suggested that the role of the board should be confined to establishing intelligence priorities, with a coordinator functioning under the jurisdiction of the Joint Chiefs of Staff. The JCS would handle the allocation of cryptanalytic tasks and intercept coverage.[28]

On 3 February 1948 the board tentatively defined its role as being one of providing "authoritative coordination" rather than "unified direction."[29] In essence, the board adopted a modified Army and Navy position. As established under the earlier Army-Navy operating agreement, there would be a Coordinator of Joint Operations, but he would continue to operate under the direction of USCIB rather than JCS.[30]

At the same meeting, USCIB accepted a draft executive order and a draft revision of its charter. The board then established a rigorous method for approval. It stipulated that the draft documents be forwarded to the chief of intelligence of each member organization for discussion. The DCI had the responsibility for clearing the documents with the secretary of defense. When everyone had approved the drafts, a master memorandum would be signed by the secretary of state, secretary of defense, secretary of the army, secretary of the navy, and secretary of the air force. This memorandum, in turn, would then be transmitted to the DCI for presentation to Admiral Leahy, who in turn was to present it to the president.[31]

In his 13 February 1948 response to the three service secretaries, Secretary of Defense Forrestal dramatically rejected the proposed documents. He felt strongly that the use of an executive order was not necessary and that in accordance with the National Security Act of 1947, USCIB should be subordinate to the National Security Council and that the NSC was the proper office to provide direction to the board.[32] His memorandum caused considerable anguish within the military structure as the JCS, Army, and Air Force expressed a strong disagreement with the direction of the guidance. They believed that a military structure and not a joint board such as NSC should be the primary voice in the control of USCIB activities. The Navy, however, did not have any problem with the proposal. From the outset of the discussions, the Navy viewed COMINT as a national resource and endorsed the placement of USCIB under the NSC.[33]

Before the dispute was finally settled, Forrestal and Hillenkoetter played major roles in shaping USCIB's new charter. Hillenkoetter pressed for a more important role for CIA, for CIA control of the COMINT function, and for the placement of USCIB under the National Security Council. Forrestal, despite his role as the secretary of defense, also clearly preferred that the USCIB not be placed in a subordinate position to the military structure. He consistently advocated that there should be some sort of a direct relationship between USCIB and the NSC. He believed that these were national functions and therefore demanded a correlation with a national authority representing all elements of the government. Forrestal's concepts on this point paralleled the Navy's views as well as those of Admiral Souers, the first DCI, now serving as the executive secretary of the National Security Council.

Forrestal's opposition to the use of an executive order for promulgation of the USCIB charter apparently stemmed in part from earlier instructions he had received from President Truman. Truman wanted to minimize the number of requests for presidential orders on the premise that the establishment of the National Security Council had as one of its main purposes the removal of this onus from the chief executive.[34]

Irritated over the lack of progress in issuing USCIB's charter, the secretary of defense addressed a second and final memorandum to Hillenkoetter on 17 March 1948. Forrestal stressed that he had not changed his mind since his original memorandum and that the DCI should get on with the business of providing uscm

with its new charter. Forrestal's main objective was to achieve the placement of USCIB under the NSC. The wording of Forrestal's memorandum left no doubt that the charter would ultimately appear in one of the regular National Security Council Intelligence Directives (NSCID).[35]

With Forrestal's memorandum, the level and intensity of the debate changed. Assured of the backing of Forrestal, CIA redrafted the charter to give the DCI greatly expanded authority and control. This second draft would have established USCIB under the NSC to effect "authoritative coordination and unified direction" of COMINT activities and to advise the DCI in matters relating to the protection of COMINT sources and "those matters in the field of communication intelligence for which he is now responsible or may hereafter be responsible."[36] When the Navy member objected to the DCI exercising "unified direction" over COMINT activities and to the broad extension of the DCI role to "matters for which he may hereafter be made responsible," Hillenkoetter acquiesced and modified his draft.[37]

His new proposal established USCIB under the NSC and authorized it to act for the NSC, under the principle of unanimity (except in electing a chairman by majority vote), in carrying out its responsibility for "authoritative coordination" (but not "unified direction") of COMINT activities. It also stated that the board would advise the DCI in "those matters in the field of COMINT for which he is responsible."[38] At the thirty-first meeting of USCIB on 13 May 1948, USCIB approved the redraft with only minor editorial amendments.[39] On 18 May 1948, the DCI forwarded the proposed directive to the executive secretary of the NSC.

But the struggle was far from being over. Admiral Sidney W. Souers, executive secretary, NSC, returned the draft to Hillenkoetter for reconsideration and further discussion. Souers, who had served as the first DCI, suggested the strengthening of the role and authority of the DCI. In particular, he sought to establish the DCI as the predominant authority in USCIB.[40]

After receiving the comments from the NSC, Hillenkoetter withdrew his concurrence of the earlier USCIB version. Hillenkoetter now had two powerful supporters – Forrestal, who insisted that USCIB be subordinated to the NSC, and Souers, the former DCI, who urged that the role of the DCI should be greatly strengthened in terms of its relationship with USCIB. Combining these elements, Hillenkoetter developed a third and final draft that would dramatically change the nature of USCIB.[41]

A central element of this revision was the downgrading of USCIB to serve in the role of advisor and assistant to the DCI. As the designated agent of the NSC, the DCI would now become not only the coordinator of COMINT activities, but the overseer as well. In the process, the DCI would also acquire the responsibility for executing all NSC directives. In short, the revision reflected a substantive change of policy as it would change the nature of the USCIB structure, and it would give to a non-military agency a position of coordination and control in the field of military departmental intelligence.[42]

Because of Hillenkoetter's reliance on Forrestal, the service intelligence chiefs felt powerless about the new developments. The Army and Air Force were particularly upset with the concept of CIA's controlling the shaping of the charter and the attachment of USCIB to the National Security Council. Although the service intelligence chiefs strongly opposed the new charter draft, they failed to make any headway against it. The civilian service secretaries (Army, Kenneth C. Royall; Navy, John L. Sullivan; Air Force, W. Stuart Symington) were reluctant to confront Forrestal after he had stated his views of the USCIB charter so forcefully and unequivocally. [43]

Hillenkoetter's third draft, which greatly strengthened the DCI role over the board, went too far. It evoked strong opposition from almost every-

W. Park Armstrong, Jr.

one. The Department of State, in the person of W. Park Armstrong, Jr., became the spokesman for this new opposition. In a memorandum to all the members of USCIB of 7 June 1948, Armstrong protested that the draft reverted to a viewpoint previously considered objectionable by the Department of State and that "it contravened, misconstrued, and overlooked many fundamental principles requisite to a secure and efficient utilization of communications intelligence by the United States Government." Armstrong objected to the DCI becoming the national authority and coordinator for COMINT activities. He also argued against the parallel downgrading of USCIB to the status of a mere advisory mechanism, existing solely for the benefit or the DCI. He maintained further that no consideration ever justified giving the DCI a position of primary control over departmental intelligence. Armstrong stressed that it was unnecessary under the law and militarily unsound to place the head of a non-military agency in control of strictly military functions that were vitally important to the military departments and that were integrated within the military commands. Armstrong's memorandum insisted on the reinstatement of the earlier version in the second draft of 13 May 1948 as unanimously accepted by USCIB and the Intelligence Advisory Council (IAC).[44]

In retrospect, the role played by Armstrong on the USCIB was highly unusual. Armstrong's position was contrary to the normal pattern in which the State Department usually joined forces with the CIA in opposition to the military control of COMINT activities. Although the State Department was indeed generally unhappy with the military direction of COMINT activities, it now became alarmed about the obvious CIA ambitions to acquire direct control of all intelligence. This resulted in Armstrong's forming a most unusual alliance with the military members in opposition to the DCI. State's position was quite clear. If USCIB needed a new national charter, State preferred the status quo to DCI domination. Major changes in the military structure would have to come later.

Following the vigorous protest by Armstrong, Hillenkoetter, as directed by USCIB, submitted a summary report of all viewpoints on 7 June 1948 to the executive secretary of the NSC.[45] With the secretary of state now a member of the NSC, and with the other members of USCIB unanimously opposed to the third revision, it became clear that the DCI (who was not a member of the NSC) had lost the battle. In its deliberations, the NSC approved the earlier version endorsed by USCIB.

The National Security Council on 1 July 1948 issued NSCID No.9, "Communications Intelligence." This directive was a major organizational turning point for USCIB and the newly established DCI. USCIB now had an official national charter that linked its subordination to the National Security Council. The charter also reflected another significant change: It accorded a new status and prominence to the agencies on board, namely, the DCI and the Department of State.[46]

Since the FBI had voluntarily dropped out of USCIB earlier, the new board was reorganized with two representatives from each armed service, two representatives from the , and two from the CIA. Except for the election of a chairman based on majority vote, decisions of USCIB continued to require the unanimous vote of all members. Other than establishing a basic change in its

subordination, the new charter did not strengthen the authorities and responsibilities of USCIB. USCIB had "authoritative coordination" not "direction or control" over all COMINT activities.

In summary, the period 1946-1949 marked the beginning of efforts to establish a new central mechanism for the handling of national security matters, including unification of the armed forces. With President Truman serving as a focal point, these efforts resulted in the enactment of the National Security Act of 1947, which created a new hierarchy for the handling of national security and intelligence matters. The act established a National Security Council to serve in an advisory capacity to the president, and a Central Intelligence Agency to be headed by a director of Central Intelligence, who reported to the president. The act also directed a major realignment of the military structure.

The first impact of these new authorities on the COMINT world took place in USCIB. DCI Hillenkoetter, with strong support from Secretary Forrestal, succeeded in obtaining a revision of the charter for USCIB. On 1 July 1948 the National Security Council issued National Security Council Intelligence Directive Number 9, a new charter for USCIB that subordinated USCIB to the NSC rather than to the military authorities. This provided positive recognition to the growing concept that intelligence matters had broader ramifications than those of purely military connotations or interests. The civil agencies, such as State Department and the CIA, now had a new status and voice in a forum previously dominated by the military organizations.

Forrestal supported Hillenkoetter on the subordination issue, but his support ended there. In his eagerness to revise and expand the charter, Hillenkoetter alienated the military and State Department representatives in his efforts to acquire substantive CIA control over COMINT, as well as over other intelligence matters. As a result, a very superficial and limited revision of the charter emerged, which meant that many of the vestiges of military control remained. Among other limitations, the revision perpetuated the requirement for unanimity among the membership on issues passed to the Board for decision.

On the military side, the National Security Act had an additional impact on the COMINT community with the establishment of the Air Force as a separate department. This resulted in the establishment of the new COMINT organization, the U.S. Air Force Security Service. A third military COMINT organization now existed competing for COMINT resources and for the assignment of cryptanalytic targets and tasks.

Chapter IV

Creation of the Armed Forces Security Agency, 1949-1952

On 20 May 1949 Louis A. Johnson, secretary of defense, established a new defense agency, the Armed Forces Security Agency (AFSA). Johnson directed a merger of the COMINT processing activities of the Army and Navy and placed the new AFSA structure under the control and direction of the Joint Chiefs of Staff. AFSA's mission was to conduct all communications intelligence and communications security activities within the Department of Defense, except those performed by the military services. With AFSA, Johnson hoped to achieve a degree of unification of the services as well as "efficiency and economy" in the management of the cryptologic structure. He also sought to minimize the resource and duplication problems associated with the new Air Force Security Service and its rapidly expanding cryptologic organization.

Louis A. Johnson, secretary of defense

The predominantly negative reactions to AFSA included the usual controversy over unification, as well as jurisdictional concerns over basic intelligence authorities and relationships. The Navy and Air Force had opposed consolidation, while the Army supported the general concept. Even greater protests came from the USCIB structure, mainly from the representatives of the Department of State and the CIA. They maintained that the AFSA charter was in direct conflict with the new national charter of USCIB. Johnson's refusal to consult or even to coordinate with USICB added to the ill will. Elements in the JCS moved to modify AFSA's charter within a few months after its establishment. JCS 2010/6 accomplished a number of substantive

changes in the AFSA charter, all of which sought to weaken the role and authorities of AFSA. The changes stressed the autonomous role of each military COMINT organization. The net result of these actions was not unification, but an acceleration of the controversy within the intelligence community over the control of the COMINT structure.

AFSA failed for two reasons. It did not succeed in centralizing the direction of the COMINT effort; and it largely ignored the interested civilian agencies – the Department of State, the CIA, and the FBI. The director did not have the authority to "direct" the military services, nor did he have the authority to suppress conflicts and duplication among the Army, Navy, and Air Force. As a result, AFSA spent most of its existence negotiating with the services over what it could do. The full extent and impact of the operational weaknesses of AFSA did not become widely recognized until the beginning of the Korean War in June 1950.

As part of the original National Security Act, Congress created the United States Air Force on 18 September 1947. The separation of the Air Force from the Army resulted in an additional cryptologic branch, first through the Air Force Security Group (AFSG) and then through the U.S. Air Force Security Service (AFSS). In the transition to a three-service structure, the Army continued to provide the Air Force with support, including cryptologic activities. On 3 June 1948, Headquarters, ASA, established the AFSG (as part of its Plans and Operations staff), with an initial cadre of eleven officers under Major Idris J. Jones.[1] The Air Force, however, clearly preferred to have direct control of its own COMINT production. On 20 October 1948 Air Force officials established a new major Air

Force command, the Air Force Security Service (AFSS), with temporary headquarters at Arlington Hall Station.

Over the next three years the Air Force, as an independent service, had a major impact on the Army-Navy COMINT organizations. The establishment of the AFSS indicated that Air Force officials demanded a security service or cryptologic organization equal to the Army or the Navy COMINT structures.

Competition for COMINT targets and resources intensified. Moreover, the three services now competed for recruitment of personnel to augment (or replace) the dwindling military competence in COMINT.

Traditionally, the Army and the Navy held divergent views on the use of civilian personnel in manning their COMINT organizations. Reliance on civilian personnel represented no major change for the Army. Throughout the 1930s the Army recruited high-level mathematicians as well as civilians with other technical skills for its cryptologic organization. During World War II, the Army continued to rely on a large civilian workforce. In contrast, at the conclusion of the war, the Navy COMINT organization employed only about seventy civilians out of almost eleven thousand persons.[2] The latter, however, did include a large number of engineers, scientists, and college professors who became naval officers during the war. Many of these were recruited personally by Commander Joseph N. Wenger from 1938 to 1939, as part of his attempt to seek candidates for the Naval Reserve program and cryptologic work.[3] Wenger's ultimate purpose, of course, was to offer direct commissions to selected individuals, thereby facilitating their immediate assignment to the Navy COMINT organization. This situation reflected the traditional Navy policy of having a purely military organization. The Navy philosophy stressed the belief that a military organization permitted stricter security control, facilitated rotation of personnel, provided greater flexibility in assignments, and reflected better overall control of COMINT operations by the military commanders.

In the postwar period of budget reduction, Navy officials found it necessary, however, to modify drastically their deeply ingrained opposition to the use of civilians in their COMINT organization. However desirable a completely military organization might be in war, these Navy officials recognized that such an approach would be difficult to achieve in peacetime. They therefore established a number of positions in the upper levels of the civil service grades for key cryptologic personnel. For many policy makers, this raised the old fundamental question for possible conflict between military and civilian authority. The issue became a matter of reconciling how naval and civilian authority could exist side by side and still retain naval control over the COMINT mission.

In effecting the transition to a greater use of civilian personnel, the Navy rationalized the change on the basis of a definition concerning the two types of control, "management" and "technical."[4] It defined management control as the day-to-day administration and control of operations of a unit in the performance of its primary function. It defined technical control as the specialized or professional guidance needed by a unit to perform its primary function. Using this distinction between management mechanisms, Navy officials in the postwar era delegated responsibility for technical direction over COMINT matters to top-level civilian employees, while reserving management and policy authorities exclusively for military officers. Under this concept, the Navy authorities determined "what is to be done while the civilian technicians determined how it is to be done."[5]

The Navy's change of heart about the use of civilians in COMINT activities and its rationalization about the nature of their role generally resolved the Navy dilemma. But manpower problems still remained for the three services. As the new Air Force Security Service sought to recruit personnel from the Army and Navy COMINT organizations, competition between the services for the dwindling manpower intensified. This situation caused Department of Defense authorities to focus anew on the basic question of whether a consolidated and centralized agency should be established. In addition to manpower questions, there existed a number of other operational considerations as well. For example, on the technical side, since the cryptography of the Soviet Union was known to be centrally controlled, a growing recognition developed among U.S. policy makers that a centralized cryptanalytic attack by the United States on Soviet systems would be beneficial.[6]

In late July 1948 Kenneth C. Royall, secretary of the army, formally brought the problem to the attention of Forrestal. Royall reasoned that the only way to avoid the increased costs associated with the new AFSS would be to establish some form of unified or joint security agency capable of serving the armed forces as a whole at the Washington level. He noted that field COMINT and security functions should probably continue to be the responsibility of the separate service departments and recommended that the secretary establish a study group to review the entire question of unification of the COMINT effort.[7]

By this time, "economy and efficiency" had become the watchwords of the Truman administration as it sought to balance the national budget. A major corollary to this recurring budgetary theme was the continuing campaign to reorganize or unify the armed forces, especially their intelligence apparatus. This issue represented a carry over, in large part, from the earlier congressional investigations into interservice cooperation – or the lack thereof – preceding the attack on Pearl Harbor. The majority report of the congressional investigation of the Pearl Harbor attack recommended on 20 July 1946

that there be a complete integration of Army and Navy intelligence agencies in order to avoid the pitfalls of divided responsibility which experience has been

made so abundantly apparent; . . . efficient intelligence services are just as essential in time of peace as in war.[8]

Drawing on this report, Forrestal and his advisers took a critical look at the military COMINT services. They questioned the wisdom of having a dual track Army-Navy COMINT structure, especially given the complexity and centralization of the Soviet communications security effort; the need for developing a coordinated United States analytic attack against the Soviet COMINT target; and the need for recruitment and training of personnel in unique cryptologic technical skills. Prompted by the specific proposal from Royall, and by the skyrocketing costs for cryptologic activities, Forrestal was looking for a way to avoid an increase in the budget. He associated the projected higher costs for cryptologic activities primarily with the plans of the Air Force for a new COMINT agency of its own. Hoping to reduce COMINT costs by preventing duplication from becoming triplication, Forrestal decided to postpone the Air Force plans for expansion until he had explored the feasibility of combining all military COMINT activities at the Washington level.[9]

Forrestal referred Royall's memorandum to his War Council, the new advisory mechanism established under the National Security Act of 1947, whose mission was "to advise the Secretary of Defense on matters of broad policy pertaining to the Armed Forces."[11] Chaired by Forrestal, the War Council was composed of the three service secretaries and the service chiefs. On 3 August 1948 the council recommended that a study group be established to consider the cryptologic needs of the entire government, including both military and civilian interests.

Accepting the council's recommendation, on 19 August 1948 Forrestal established a military committee to consider the "creation of a Unified Armed Forces Security Agency. The Terms of Reference for the committee gave the study a purpose, directed at both foreign communications intelligence and the security of United States communications.[11] Forrestal's general mandate to the committee considered two broad questions:

Should there be created a joint or unified Armed Forces agency for the production of communications intelligence and, if so, what form should it take?

Should there be joint or unified cryptographic security activities of the Army, Navy, and Air Force, and if so, what form should they take?

The committee consisted of six officers: for the Army, Major General Alexander R. Bolling, Assistant Chief of Staff, G-2, and Colonel Harold C. Hayes, Chief, Army Security Agency; for the Navy, Rear Admiral Earl E. Stone, Director, Naval Communications, and Captain William S. Veeder; and for the Air Force, Major General Charles P. Cabell, Director of Intelligence, and Brigadier General Francis L. Ankenbrandt, Director for Communications.[12]

At its first meeting on 25 August 1948, the committee selected Admiral Stone as its chairman. This resulted in the group's common designation as the Stone Board.[13] Hayes, however, was the only committee member actively engaged in the production of COMINT at the time, as well as the only member having any experience in COMINT.

While the committee focused essentially on communications intelligence and communications security activities from a military point of view, it also recognized the cryptologic interests of other parts of the government. Its Terms of Reference instructed it to consult with the State Department and the CIA as part of its fact-finding effort for the preparation of a final report.[14]

Despite Forrestal's interest in consolidation, the establishment of the new agency did not come about easily. Lasting several months, the Stone Board deliberations revealed that the Navy and the Air Force were not ready to accept the kind of unification proposed by the Army. After considerable debate, the Stone Board submitted its final report to Forrestal in December 1948.[15] The Stone Board's report, actually a majority report and an accompanying minority report, did not reconcile the conflicting views of the services.

The majority report, written by the Navy and Air Force, essentially recommended a continuation of the basic arrangements existing under the old Joint Operating Plan. It added several organizational changes related primarily to the new Air Force Security Service and proposed exempting tactical support areas from central control. The majority report also proposed the allocation of joint tasks to a new Air Force agency "on an equal basis with the Army and Navy." In addition, the Navy and Air Force sought to exempt from unified control those parts of the effort that pertained directly to the specific military responsibility of each service and that would remain in each service as a command function. Navy and Air Force officials justified their recommendation in terms of the need for flexibility of operation and speed of decision in matters pertaining to responsibilities of the services during wartime. In short, they basically desired to maintain the status quo and their own separate and independent cryptologic efforts. Consolidation was not in their plans.

In the minority report, the Army officials took a far different approach. They emphasized economy and the avoidance of duplication of effort.[16] The Army officials argued that foreign communications were becoming increasingly sophisticated and consequently were much more difficult to exploit. Accordingly, the Army plan placed primary importance on maintaining and exploiting technical relationships among all analytic problems. To achieve this capability, the Army recommended consolidation of the COMINT services into a single unified agency. The unified agency would have responsibility for central control of processing and dissemination activities for the entire U.S. COMINT effort. In effect, the minority Army plan proposed that all

COMINT production other than intercept and field processing be conducted by one organization, staffed by personnel from the three services. This new unified agency "would determine COMINT implementation priorities based on intelligence requirements. . . and would determine the specific employment of the intercept facilities."[17]

Under the Army concept, no single service would perform central processing activities in the United States. The services could, however, perform a limited field processing effort, primarily on tasks of a direct support nature as necessary for military operations of each service. The individual services would each maintain COMINT organizations to conduct intercept, direction finding, and necessary field processing; to train service personnel; and to engage in research and development actions for COMINT operations. As part of their basic responsibility to the new joint agency, the services would provide "intercept facilities and personnel." Their fixed field sites would conduct intercept activities in response to the tasking of the new joint structure.

In support of its proposal, the Army drew a parallel between the Navy-Air Force proposal and the situation that prevailed in the German COMINT services during World War II. According to the Army, at the end of the war all of the German COMINT services and agencies were independently duplicating the work of one another in an atmosphere of great hostility. The Army believed that the establishment of three independent COMINT activities would be divisive to United States interests and would, in time, degenerate into a situation similar to that encountered by the German SIGINT services during the war.[18]

In contrast to the German structure, the Army cited the coordination and accomplishments achieved by the British in their intelligence structure. In the Army's view, the COMINT organization of the British, with approximately thirty-five years of continuous experience in this field, had been unified in a manner similar to that described in the Army proposal.[19] Citing the German and British examples, the Army plan stressed the need for consolidating into a single armed forces activity all but the most narrowly defined problems of primary interest to each service.

While the majority and minority reports disagreed on the issue of consolidation, they did agree, in principle, on the need for better integration of overall COMINT support. Both plans agreed that all three services should participate in this integration under the coordination of an Armed Forces Communications Intelligence Board (AFCIB) and a Director of Joint Operations (DJO).[20] Both reports proposed that the new Armed Forces Communication Intelligence Board be subordinate to the Joint Chiefs of Staff and that it consist of the military members of USCIB. This new AFCIB would be the policy board for providing guidance to COMINT activities, and would provide liaison with USCIB on all matters within the cognizance of USCIB. The role of the DJO would vary, however, under each report. Under the majority report, the DJO would represent an expanded role for the existing Coordinator of Joint Operations. Under the Army report, the DJO would, in effect, have two hats, serving as director of the new unified agency, and as the Director of Joint Operations.

The military services were not the only ones interested in the Stone Board Report. From the outset, the civilian agencies – the Department of State and the CIA – followed the activities of the Stone Board closely, and soon expressed strong objections to various parts of the final report. The main issue raised by the CIA and the State Department concerned the establishment of an Armed Forces Communications Intelligence Board whose relationship to USCIB was unclear. CIA and State authorities viewed the new board as a threat to USCIB.[21]

While the relationships of AFCIB may have been unclear, the strategy of the military services was obvious. Their intent was to establish AFCIB as a purely military structure running parallel to

USCIB, but as one that would contravene the policy role of USCIB. It would, in effect, leave USCIB with no significant role. With this proposal, the military had resurrected the acrimonious issue of one year earlier when the NSC had established USCIB as subordinate to the NSC rather than the JCS.

Rear Admiral Roscoe H. Hillenkoetter, , strongly argued that the creation of an Armed Forces Intelligence Board would clearly be in conflict with the new authority granted to USCIB by the issuance of NSCID No.9 in 1948.[22] In his view, the creation of an AFCIB would serve to give the military authorities total control over activities that should, in the national interest, be directed more to the requirements of the civil agencies. Hillenkoetter also took the position that in the Cold War the Central Intelligence Agency and State Department were the primary players on the covert and diplomatic front. According to his estimate, three-fourths of the current production of COMINT came from sources of primary interest to the CIA and the State Department.

W. Park Armstrong, special assistant to the secretary of state, also attacked the concept of an AFCIB as a "military controlled structure parallel to USCIB and independent of it." His strong objection also reflected the view that the non-military consumer would lose the ability to influence the military COMINT structure in terms of stating its requirements for COMINT information.[23]

The civilian objections, along with the basic service disagreement over consolidation, prevented any immediate action. A serious illness suffered by Forrestal also contributed to the postponement of any final decision. The two reports submitted by the Stone Board awaited disposition in the office of the Secretary of Defense for more than four months.

When Louis A. Johnson became the secretary of defense on 28 March 1949 following Forrestal's death, he acted quickly to resolve the issue. He called in General Joseph T. McNarney, USA, known to be a supporter of the consolidation concept, to assist in resolving the dilemma.[24] McNarney, who served as director of Management Services for Johnson, recommended a plan that required a merger, but left the three services the right to maintain their separate organizations. It was a compromise solution. Johnson later reissued it as a draft directive calling for the establishment of an Armed Forces Security Agency that would be along the lines of the recommendations in the Army minority report. The new Johnson directive was then scheduled for discussion and decision by the JCS and the Secretary's War Council.

At a JCS meeting in the morning of 18 May 1949, General Hoyt S. Vandenberg, Air Force chief of staff, suddenly announced the reversal of the earlier Air Force position and indicated that the "Air Force supported the Army's consolidated concept and not the Navy's non-consolidated concept for COMINT processing."[25] Vandenberg appears to have voted for merger only after having obtained prior assurance that each service would be allowed to have its own agency for the conduct of those cryptologic operations peculiar to its needs.[26] With this official reversal of the Air Force vote, the Navy remained the sole dissenter to the establishment of an Armed Forces Security Agency. But its position soon collapsed. At an afternoon meeting on the same day, the Secretary's War Council met to consider Johnson's draft directive. At the council meeting, the newly appointed secretary of the navy, Francis P. Matthews, overruled the position of the Chief of Naval Operations, Admiral Louis Denfield, and argued for consolidation of the COMINT structure.[27] Thus, by the evening of 18 May Johnson had succeeded in overturning the split report of the Stone Board and gaining the support of all three services for consolidation.

The reasons for the change of position by the Air Force and Navy officials were probably associated more with high-level political factors rather than any conceptual changes by the services themselves. From 1947 to early 1950, President Truman, Congress, and the secretaries of defense actively supported a concept of genuine unification of the military services. As a result, this period was characterized by a number of bitter interservice rivalries and disputes concerning the issue of unification as well as questions concerning the role of each service. Many of the same officials who participated in the Stone Board decision were also active in other ongoing political battles within the National Military Establishment. One such battle, associated with the Navy's desire to hold nuclear weapons, came to a climax at the time of the deliberations about AFSA in 1949. The Navy and Air Force were in violent disagreement over the strategic role of the Navy and clashed over the construction of a new super carrier and the development of the latest strategic bomber, the B-36. This resulted in a chain reaction of events achieving national prominence, and soon involved both the executive and legislative branches of the government. These conflicts resulted in Johnson's cancellation, on 23 April 1949, of the construction of the 65,000-ton aircraft carrier, the USS *United States*; the resignation of Secretary of the Navy John L. Sullivan on 19 April 1949; and the continuing public rivalry between the Air Force and Navy, usually labeled "the revolt of the admirals."[28]

Thus, the atmosphere for cooperation and consolidation was volatile. However, by the time the vote on the AFSA merger took place, Johnson had appointed a new secretary of the navy, Francis P. Matthews, whom he had selected personally.[29] Matthews supported Johnson's position. As for the Air Force, Vandenberg, as Air Force chief of staff, had consistently and actively promoted the growth of the Air Force as an independent service. In addition to the commitment for an independent cryptologic arm for the Air Force, it is also likely that Vandenberg envisaged receiving greater financial support for aircraft programs if he went along with consolidation in other areas such as the merger of cryptologic activities.

There also existed other pressures on the services. As a political reality, the Navy-Air Force decision to vote for merger was consistent with President Truman's desire for unification of the military COMINT organizations which, in turn, reflected the mood in Congress, as well as the tenor of the National Security Act. In later years, Brigadier General Carter W. Clarke, USA, a participant in the AFSA deliberations, commented about the political pressures existing at the time. Clarke, an official with many years of service in ASA and the office of the Army Chief of Staff for Intelligence, maintained that while Johnson pressed hard for merger, the real leverage for final service approval of the AFSA concept came from Truman and Congress.[30]

During the sessions of the Stone Board, the Navy made elaborate studies and charts of the major elements in both the majority and minority reports. As a result of this critical analysis, the Navy developed a unique grasp of the weaknesses and strengths of each report. Recognizing that the establishment of a consolidated agency was inevitable, and seeking to strengthen the operations of the new agency, the Navy sought at the eleventh hour to have some positive changes made to strengthen the new organization.

On behalf of the Navy organization, Stone conferred at length with General McNarney about the directive and proposed some substantive changes in the text. In particular, the Navy sought changes in the charter concerning the distinction existing between "fixed" and "mobile" collection sites. The draft directive stated that the fixed intercept installations would be "manned and administered by the service providing them, but will be operationally directed by AFSA." On the other hand, the mobile sites were to be excluded from any AFSA control and would be operationally controlled by the parent service. With unusual foresight, the Navy sought to eliminate this distinction, which it correctly predicted would be an issue of great difficulty for the new agency. McNarney, however, refused to make any substantive changes, insisting that his authority, as executive secretary of the War Council, extended only to editorial changes.[31]

The Johnson Team
left to right: Secretary of the Army Gordon Gray; Secretary of Defense Louis A. Johnson; Secretary of the Navy Francis P. Matthews; and Secretary of the Air Force Stuart Symington

With a unanimous vote now supporting the concept of merger, Johnson was ready to act. On 20 May 1949, Johnson ordered the issuance of JCS Directive 2010, establishing an Armed Forces Security Agency for the conduct of communications intelligence and communication security activities within the National Military Establishment. The new agency would be headed for a two-year term by a flag or general officer, to be chosen in turn from each service. The agency would function under the management control of the JCS, and would conduct its common activities "in not more than two major establishments." Johnson's directive established a date of 1 January 1950 for completion of the merger of the COMINT services.[32]

In taking this decisive action only two months after coming into office, Johnson ended the impasse that had existed for over a year. With the formation of AFSA, the military COMINT structure acquired a new identity and structure. While Johnson sought to recognize the unity of the COMINT mission and resources, he believed that a consolidation of the service COMINT efforts would be responsive at the same time to the public pressure for effecting greater economies in government.

Because of the concerns expressed earlier by the civilian agencies, Johnson also sent parallel letters to Dean Acheson, secretary of state; Rear Admiral Roscoe H. Hillenkoetter, director of Central Intelligence; and Rear Admiral Sidney W. Souers, executive secretary of the National Security Council.[33] He informed them of the formation, with the approval of the president, of a unified cryptologic establishment – the Armed Forces Security Agency (AFSA) – and of the subordination and missions of the new structure. The tone of the letters was conciliatory as Johnson sought to assure each that the implementation of his new directive would not interfere with the functions of the USCIB. The final AFSA charter included no vestiges of a high-level Armed Forces Communications Intelligence Board, which had drawn such violent protests earlier from the CIA and the Department of State.

The JCS then reissued Johnson's directive as JCS 2010 on 20 May 1949. The JCS established a steering committee to assist in the planning of the many administrative, logistic, and operational actions necessary for physical merger of service resources. Its members were those representatives of the military departments then serving on USCIB.

They were, for the Army, Major General S. Leroy Irwin and Colonel Carter W. Clarke; for the Navy, Rear Admiral Thomas B. Inglis and Rear Admiral Earl E. Stone; and for the Air Force, Major General Charles P. Cabell and Colonel Roy H. Lynn.[34]

In the selection process for the position of the director, AFSA, each service proposed one candidate. The nominees were Major General J. V. Matejka, USA, Chief Signal Officer, USEUCOM; Rear Admiral Earl E. Stone, USN, Director of Naval Communications; and Major General Walter E. Todd, USAF, Joint Staff. On 15 June 1949, the JCS selected Stone.

The appointment of Stone, who had no experience in COMINT, was a sign to some that consolidation might be aborted. Stone, who represented the service most consistent in its desire for cryptologic autonomy, had signed the majority report opposing the creation of the agency he was now to head. Consolidation seemed in jeopardy.

Rear Admiral Earl E. Stone, first director of AFSA

As a result of AFSA's creation, the existing CJO (Admiral Wenger) and the director, AFSA, now had overlapping responsibilities. The CJO had been established earlier as USCIB's executive for the discharge of certain responsibilities, which the services, along with the civil agencies, had agreed should be vested in USCIB. Subsequently, with the establishment of AFSA and without prior concurrence of USCIB, these responsibilities were arbitrarily assigned to the director, AFSA. This situation presented a major dilemma for USCIB, particularly in the policy area of foreign liaison, a responsibility that NSCID No.9 assigned to USCIB.

At the forty-first meeting of USCIB on 17 June 1949, the representatives of the Central Intelligence Agency (Admiral Hillenkoetter) and the Department of State (W. Park Armstrong) cited the conflicts existing between the new AFSA directive and the responsibilities of USCIB and the role of the CJO under NSCID No.9. The military members of USCIB were unresponsive to the complaints of the CIA and State Department members, refusing essentially to go counter to a directive from the secretary of defense. The meeting ended with the agreement that the CIA and State Department should address a letter to Secretary Johnson expressing their views about the conflicts between the USCIB and AFSA charters.[35]

On 27 June 1949 Hillenkoetter sent such a letter to Johnson. Johnson, however, took no action. General McNarney, head of Johnson's management committee, met with USCIB to discuss the letter, but adamantly refused to pursue any changes to the AFSA directive.[36] Hillenkoetter and Armstrong concluded that McNarney's views were inflexible and that further effort to change them was futile. They would have to wait for a new opportunity.

Although USCIB's non-military members were unanimous in their opposition to the establishment of AFSA, they were unable to convince USCIB to take any stronger action protesting the establishment of AFSA. Because of charter limitations and the preponderance of military membership, USCIB would not challenge Johnson's plan. Recognizing that they could do nothing about the establishment or the structure of AFSA, the USCIB had no choice but to cooperate with the new agency during the interim period of its establishment. Further, lacking any response to Hillenkoetter's letter, the only solution for USCIB out of its jurisdictional dilemma was to appoint the director, AFSA, as the CJO.[37] In his capacity as CJO, the director of AFSA had a parallel responsibility to USCIB, which made the situation salvageable for USCIB. Consequently, USCIB agreed that the existing CJO (Admiral Wenger) should serve as the deputy

CJO during the transition period, to assist primarily in the area of Second Party collaboration.

When Johnson established AFSA in May 1949, he simultaneously ordered the establishment of an advisory mechanism to exist within the AFSA structure. His directive defined the nature, role, and composition of the Armed Forces Communications Intelligence Advisory Council (AFCIAC), designed to serve solely in the capacity of an advisory mechanism for AFSA and the JCS. Johnson's directive, however, indicated very clearly that AFSA, subject to the JCS, had the primary responsibility for formulating and implementing plans, policies, and doctrine relating to communications intelligence and communications security activities. AFSA, in effect, had the actual responsibility for running the COMINT and COMSEC operations, excluding only those responsibilities that were delegated individually to the Army, Navy, and Air Force.[38]

This dichotomy of organizational roles and division of labor did not last long. Within two months after the establishment of AFSA, the JCS proposed substantive changes to Johnson's directive. On 28 July 1949 JCS issued a new charter for AFCIAC, which not only changed the nature and role of the AFCIAC mechanism, but affected areas of authority of AFSA as well. After circulating the document to Johnson for information, the JCS approved the new AFCIAC charter on 1 September 1949.[39]

The changes in JCS 2010/6 tended to diminish the connotation of AFSA as representing a "unified" organization, while at the same time placing greater emphasis on "joint" operations. JCS accomplished this by transferring responsibilities to AFCIAC from the JCS, as well as from the director, AFSA. As noted above, AFSA initially had the authority to formulate and, after JCS approval, to implement plans, policies, and doctrine relating to communications intelligence and communications security activities. Under the new AFCIAC charter, however, AFSA no longer performed these basic tasks. Instead AFCIAC became the structure to "determine policies, operating plans and doctrines for the

AFSA in its production of communications intelligence, and in its conduct of communications security activities." In addition to its originally assigned advisory functions, AFCIAC acquired considerable staff authority over AFSA. Thus, AFCIAC no longer functioned as an advisory mechanism within AFSA, but "became the agency of the JCS charged with the responsibility for ensuring the most effective operation of the AFSA." As the new policy authority and overseer of AFSA's operations, AFCIAC emerged as the real power in the COMINT structure.

In short, before the AFSA structure even became operational, the JCS had eroded much of AFSA's already limited authority. As happened during the period of the Joint Operating Plan, the military authorities turned once again to the use of a "committee" structure to run the COMINT organization. They also repeated the practice of excluding the civilian agencies from any participation in the committee process. These charter changes tended not only to destroy the separate identity of AFSA, but also to preserve the independent identity of the separate military COMINT services. The net result was an acceleration of the antagonism within the intelligence community over the control of the COMINT structure.

At the fourth meeting of AFCIAC, the council proposed to the JCS that its name be changed to Armed Forces Security Agency Council (AFSAC). Admiral Stone proposed the use of the word "security" because it was a generic term that embraced both the communications intelligence and communications security fields. Because of AFCIAC's jurisdiction over COMINT as well as COMSEC, the new title was considered to be more accurate. It also avoided the use of the term "communications intelligence," which at that time was considered to be classified. Following JCS approval of the change on 22 October 1949, the secretary of defense approved the change on 9 November 1949. After that, the Council was known as AFSAC.[40]

The Director, AFSA, chaired the AFSAC, which now became the mechanism through which AFSA reported to the JCS. AFSAC had ten members, including the chairman – three each from the three services, one from each of their communications, intelligence, and cryptanalytic organizations. Consistent with the traditional practice in arrangements for joint operations, actions on substantive matters brought before AFSAC could be taken only by unanimous vote. Because of the diversity of the membership and their different interests, the requirement for unanimity made decision-making difficult, if not impossible. This factor ultimately caused major problems for AFSA, as the services tended to vote along party lines whenever major issues arose. The hope that AFSA would develop a truly consolidated intelligence effort seemed remote.

Stone began his tour as Director, AFSA, and the CJO on 15 July 1949.[41] He was assisted by three deputy directors. Each deputy served as the liaison between AFSA and his parent service, and assumed specific functions within AFSA. Colonel Samuel P. Collins, the Army deputy director (AFSA-00A), was responsible for communications security, research and development, and communications. Captain Joseph N. Wenger became the Navy deputy director (AFSA-00B), and assumed control of COMINT. Wenger also received an additional duty in a major policy area. When USCIB appointed Stone as the CJO, Wenger, who was CJO at the time, agreed to step down and to become the deputy CJO. As the deputy CJO, Wenger then assumed the responsibility for directing foreign liaison relationships on behalf of the dual interests of USCIB and AFSA. On the Air Force side, Colonel Roy H. Lynn of the Air Force (AFSA-00C) became responsible for staff and administrative functions.[42]

On the question of physical consolidation of facilities, neither the Army site (Arlington Hall Station) nor the Navy site (Naval Security Station) could accommodate all of AFSA's COMINT and COMSEC functions. One proposal was to make the split along COMINT-COMSEC lines; another placed all analysis on Soviet problems at Arlington Hall and all other cryptanalytic problems (ALL) at the Navy site; a third left physical arrangements as they were, with a mixture of both COMINT and COMSEC remaining at the two installations.[43]

Stone opted for a division along the major functional lines of COMINT and COMSEC. He placed almost all of the COMINT operations and related research and development activities at Arlington Hall and all COMSEC operations, with related research and development aspects, at the Naval Security Station. Staff and other support functions were divided between the two sites depending on equipment and other logistical considerations. Stone established his new headquarters and administrative offices at the Naval Security Station.[44]

Stone also established four monitor groups – representing major functional areas – to direct the actual integration of the service elements into AFSA. The COMINT Monitor Group, headed by Captain Redfield Mason, was concerned with the merger of the largest of AFSA's operating units. This organization was to merge the Army Security Agency's Operations Division (AS-90) with the Processing Department (N-2) of the Navy's Communications Supplementary Activity, Washington (CSAW), as well as related units of both agencies. It also had the responsibility for the assimilation of two committee structures that operated under the aegis of the superseded Joint Operating Plan. This included the JPAG and the JICG of the JOP. The resulting structure became the Office of Operations, with Mason designated as its first chief.[45]

With the physical merger of the Army and Navy COMINT processing organizations into the AFSA structure, each service contributed many talented individuals with distinguished careers in cryptography. A number of these had achieved technical accomplishments of exceptional merit and possessed a combination of knowledge and experience that would become invaluable assets to the new agency. The Army's principal cryptologists were civilians, most of whom had acquired wartime com

*Captain Redfield Mason
chief, COMINT Monitor
Group*

*Dr. Solomon Kullback
chief, Research and
Development Monitor
Group*

*Dr. Abraham Sinkov
chief, Communications
Security Monitor Group*

*Captain John S. Harper
chief, Administration
Monitor Group*

*AFSA Monitor Group
Chiefs*

missions; the Navy's leading cryptologists were mostly career Navy officers.

Moving into the AFSA structure in 1949, William F. Friedman clearly stood out as the dean of Army cryptologists. During his career in the War

developed a small cadre of personnel who were highly trained in cryptology. During World War II, Rowlett, Sinkov, and Kullback, in particular, attained high personal accomplishments. They became representative of a new generation of cryptanalysts.

Frank B. Rowlett

Department and the Signal Intelligence Service, Friedman's contributions embraced not only cryptanalytic operations but research and development activities (both COMINT and COMSEC) and cryptologic education as well. While Friedman has received recognition for his abilities in the field of cryptanalysis, possibly his most lasting contribution may be in the area of training. Singularly gifted as both a teacher and a writer, Friedman left a legacy of training lectures and programs that have broad application even today. In April 1930, in his search for new talent for the newly established Signal Intelligence Service, Friedman personally recruited three young mathematicians – Frank B. Rowlett, Abraham Sinkov, and Solomon Kullback – as junior cryptologists. As the Army continued its recruitment through the 1930s, the SIS gradually

In contrast, the manpower of the Navy's COMINT organization in 1949 essentially represented a military cadre. The new Navy policy of the postwar period that moved in the direction of increased recruitment and use of civilian personnel had not been in effect long enough to have a significant impact on its overall structure. The Navy authorities, by encouraging professional careers in cryptology for its officers, had traditionally placed primary reliance on naval officers for the performance of COMINT duties. By the time AFSA was established, many Navy officers had acquired a broad background in communications intelligence matters, some dating back to the mid-1920s, including a number of assignments as either producers or consumers of intelligence. During World War II, officers certified in the cryptologic field filled positions of major responsibility for the Navy in Washington and in field installations. Some of these field installations were Fleet Radio Unit, Pearl Harbor (FRUPAC) in Hawaii; Fleet Radio Unit, Melbourne (FRUMEL) in Australia; Radio Analysis Group Forward Area (RAGFOR) on Guam; and a number of communications and intercept sites. Based on this broad depth of training and experience, AFSA acquired a talented and highly professional organization from the Navy. The list of officers assigned to AFSA with their initial assignments included Captain Joseph N. Wenger, deputy director, AFSA (for COMINT); Captain Laurance F. Safford, special assistant to the director; Captain Thomas A. Dyer, chief, Plans and Policy; Captain Redfield Mason, chief, Office of Operations; and Captain Wesley A. Wright, chief of the Special Processing Division.

The other monitor groups were Research and Development, headed by Dr. Solomon Kullback; Communications Security, headed by Dr. Abraham Sinkov; and Administration, headed by Captain John S. Harper, USN. William F. Friedman became research consultant for the new AFSA. Frank B. Rowlett, a member of the Army team that broke the Japanese diplomatic system (Purple) during World War II, became AFSA's technical director.

Initially, Stone had two roles: one as director, AFSA, and the other as the coordinator and executive agent for USCIB. As the director, Stone reported to Secretary of Defense Johnson through the JCS. He had the responsibility for all cryptologic activities in the Department of Defense. He was responsible for the furnishing of COMINT, not only to the military services, but also to other government departments. As director, he was also subject to the policies and rules of USCIB governing the production and dissemination of COMINT. However, once the COMINT was distributed to authorized intelligence recipients, Stone had no jurisdiction over its use or physical security. These became the responsibility of the user.[46] When acting as the coordinator and executive agent for USCIB, Stone worked under a different authority

and in wider fields than he did as the director, AFSA. As coordinator, Stone was responsible to the NSC through the USCIB. In this capacity, his authority extended to the use of COMINT by any U.S. agency. As director, AFSA, or as coordinator of USCIB, Stone was an ex officio member of USCIB, without a vote, and was ineligible to become chairman of USCIB (which was then held on a rotating basis among the membership).[47] Later, this lack of voting status in USCIB became a major obstacle for Stone and his successor as they were excluded from participation in the actual decision-making process of the Board.

During his first year, Stone constantly sought to clarify the nature and role of the AFSA structure and to accomplish all necessary consolidation actions. For the first seven months, his major objective was to develop administrative policies and procedures for the continuation of communications intelligence and communications security activities. He placed major emphasis on actions dealing with physical and administrative consolidation, and budget and financial factors. In his first progress report to AFSAC, Stone cited 15 July 1949 as the date of the formal activation of AFSA.[48] The report also noted the following as AFSA's milestones: On 1 October 1949, AFSA assumed operational control of its cryptologic activities; and on 25 December 1949, AFSA assumed administrative control of all its allocated civilian personnel. The transfer and consolidation of facilities and personnel in a six-month period was a significant accomplishment. Merger seemed a reality.

The contrast between the earlier JOP, and the new AFSA was readily apparent. The effort against the several targets was performed jointly by the CJO assisted by his small staff (committees) for intercept control, allocation of processing tasks, and foreign liaison. Under the JOP, the services worked together on a voluntary basis, operating essentially under a "management by committee" approach, with each service retaining its own independence – particularly when joint agreement could not be reached. AFSA differed from the JOP principally in that the COMINT processing activities of the Army and Navy in the United States were now physically merged at two locations. It performed both military and non-military tasks. It was a major step toward consolidation.

With the merger of the Army and Navy COMINT processing organizations under AFSA, there no longer existed any need for the JICG or the JPAG, both of which were creations of the earlier Joint Operating Plan. The functions of the Joint Intercept Group and the Joint Processing Group were merged into a new AFSA Office of Operations. The duties of the JLG, however, required no realignment within the AFSA structure. The JLG, which dealt essentially with foreign liaison matters, continued to be a responsibility of the CJO under his USCIB hat. The responsibility for foreign COMINT liaison, administered by the JLG and its supporting staff, remained the responsibility of USCIB. This function remained under Stone as the Coordinator of Joint Operations.

Although the processing activities of the Army and Navy were now merged and the three services now functioned under AFSA as a joint agency of the JCS, the new agency faced some fundamental problems. The services generally took full advantage of the many loopholes existing in the AFSA charter in order to preserve their independence. For example, the AFSA charter withheld from AFSA any authority for the tasking of mobile collection sites.[49] This "exclusion" clause caused serious operational problems for AFSA from the outset. Initially, the Army and Navy reserved "mobile" or close-support facilities in the field for their exclusive control. According to the military services, this ensured that each satisfied the requirements of its own commanders for the production of COMINT for tactical purposes. This blanket delegation of authority to the services proved to be a major problem for AFSA, particularly in its relationships with the newly established Air Force Security Service.

By the simple act of declaring an intercept facility as mobile, a service could withhold any collec-tion activity from Stone's control. The Air Force used this exclusion to the maximum by conveniently identifying all of its intercept facilities as mobile sites. Because of this situation, AFSA and the U.S. Air Force concluded an agreement on 18 September 1950 that essentially reflected a shared arrangement for AFSA/AFSS tasking of Air Force mobile collection sites.[50] The agreement concluded with the candid admission "that the agreement was made unilaterally between AFSA and the Air Force, in view of the fact that the latter is not providing any fixed intercept installation for operational direction of the Director, AFSA." This sharing of tasking reflected the best arrangement that AFSA was able to achieve with the AFSS. During the remainder of Stone's tour, as well as his successor's, the Air Force continued to withhold the assignment of fixed stations from AFSA's control by continuing to identify all of its sites as mobile.[51]

Later these same difficulties over the question of fixed and mobile collection sites extended to relationships with the Army, but to a lesser degree. While the Army did not exclude AFSA from tasking its mobile sites, it did establish elaborate procedural channels for the relay of AFSA tasking instructions. Except in an emergency, AFSA could not task the sites directly, but had to go through intermediary channels, such as headquarters installations in the United States and in the field. The process was cumbersome and inefficient and worked against the timeliness of COMINT reporting.[52] The Navy was the only service that did not create problems in this area.

Another broad question that plagued AFSA officials was the division of responsibility between AFSA and the services. In its relationships with the services on processing and reporting matters, for example, AFSA once again found itself at odds with the AFSS. Neither the Army nor the Navy undertook to establish processing units within the United States. The Air Force, however, insisted on having its own processing unit within the United States.[53] AFSA considered this a major violation of its responsibilities. It asked the JCS to settle the dispute. Since the JCS failed to rule publicly in favor of either organization, the Air Force finally abandoned its plans for establishing a domestic processing unit.[54] The issue, however, did not go away. During the entire period of AFSA's existence, its relationships with the Air Force remained highly contentious over this issue, as well as over the question of mobile collection facilities.

Although in principle the military cryptologic community was officially committed to making the merger work, this commitment was not reflected uniformly throughout the services. There still remained much open hostility and skepticism about the workability of the concept of consolidation. In addition, the non-military members of USCIB continued to raise questions about AFSA's role and its relationships to USCIB. They complained about the lack of a civilian voice in this military hierarchy.[55]

By June 1950 AFSA had been operational for six months. It was still preoccupied with efforts to sort out its managerial role and its authority in the COMINT structure, as well as its relationships with the consumer community. The outbreak of the Korean War on 25 June 1950, however, completely changed the focus of AFSA's activities. The war put new pressures on all U.S. intelligence sources. Both military and civilian intelligence authorities immediately pressed for an improvement in the quality and timeliness of COMINT reporting.[56] The war, however, quickly revealed the limitations of AFSA. Duplicate collection efforts, processing problems, service rivalries, and communication delays were prevalent. AFSA's limited ability to direct COMINT activities in support of national targets soon became evident to the entire intelligence community.

As the Korean War continued, the U.S. COMINT community achieved a mixed record of successes and failures. Because of the practice of counterpart coverage, each service concentrated on intercepting and processing the communications of its foreign counterpart. For example, the Army and

Air Force intercepted the communications of the Korean military ground and air forces, respectively. The Navy handled the communications of Korean naval forces. Because of this reliance on counterpart coverage, the major COMINT successes took place in the area of tactical support. The Army and Air Force, working independently on the low-grade communications of their counterpart targets, had the most success. The field exploitation of North Korean and Chinese Communist traffic both voice and plain text proved to be of significant value to the U.S. field commanders.

By late 1951, because of the continued absence of COMINT from some enemy communications, the U.S. intelligence community, both military and civilian, became increasingly impatient with the quality and timeliness of AFSA's COMINT reporting. The military desired increased expenditures of effort and personnel on the analytic problem. The civilian officials complained about the lack of channels for expressing their intelligence requirements and priorities to AFSA.[57] Pressure mounted on Stone and AFSA to improve the responsiveness of the COMINT structure.

As Stone's two-year tour was coming to a close, AFSA convened in January 1951 to nominate a successor. According to the agreed procedures, the council would consider nominations made by the Army and the Air Force, but not the Navy, since the Navy had provided the first director. The Air Force, however, declined, and supported the Army's candidate, Major General Ralph J. Canine, USA.[58] On 15 February 1951, the secretary of defense approved the appointment of Canine as director, AFSA.[59]

During World War II, Canine, an artillery officer, had served under General George Patton as chief of staff, XII Corps, Third Army. Other than having been a user of intelligence, Canine came to the AFSA job with no prior intelligence experience. However, prior to assumption of his AFSA duties, Canine had a unique opportunity to participate directly in a great number of matters involving

AFSA's responsibilities and relationships.[60] As the Army's alternate member of AFSAC and USCIB, he participated in their meetings for a six-month period and learned firsthand about many of the issues confronting AFSA. This extended period of orientation gave him a valuable preview of the AFSA structure before his formal assumption of the position.

On 15 July 1951 Canine succeeded Stone as the second director, AFSA. Canine's arrival heralded no immediate major changes in the AFSA structure, however. AFSA continued to operate under a multiple control arrangement, functioning under the guidance of USCIB and the JCS. USCIB provided limited guidance on policy matters, while the JCS provided the management and operational authority over AFSA. The Armed Forces Security Advisory Committee continued to oversee the operations of AFSA for the JCS, and at the same time exercised a heavy hand in the direction of AFSA activities. As noted earlier, AFSA did not have complete freedom of action in the policy and planning areas. very early revision of the AFSA charter required that all major policy and planning actions by AFSA had to have prior approval of AFSAC. Concerning the very critical policy question of the division of responsibility between AFSA and the services, AFSAC consistently supported the service views rather than AFSA's. AFSAC also required unanimous approval by the members prior to taking action on an issue. This meant that it was very difficult, if not impossible, for AFSA to win a favorable decision on controversial issues.

Canine's arrival also brought no real changes in terms of AFSA's working relationships with the service COMINT organizations. If anything, a steady deterioration in these working relationships continued. After the establishment of AFSA, even the Army's initial support and enthusiasm for the centralized concept began to diminish and soon matched that of the other services in opposition to unification. In the Army's case, the change was due primarily to the reassignment of Brigadier General Carter W. Clarke, USA, chief, Army Security Agency, who was considered to be the primary architect of the consolidation concept.[61] Clarke became commanding general, Southwestern Command, Japan, and remained temporarily out of the intelligence field until returning to Washington in late 1953.[62] With Clarke's departure, no one in authority in the Army wholeheartedly supported consolidation.

The services did agree on one major issue. They were united in their belief that the director should have no authority over them. They viewed him solely in the role of "coordinator," not "director." Each of the first two directors of AFSA, Stone and Canine, bitterly fought this concept. They believed this approach to be contrary to the spirit and intent of the AFSA charter. It inhibited them from doing their job properly. As a result of continuing pressure from Stone and Canine for changes, hostility continued to build between AFSA and the COMINT services.

Despite Stone and Canine's efforts, by December 1951 management and control issues remained unresolved. Critical to the successful functioning of AFSA were centralized processing; better communications, including courier forwarding of raw traffic as well as improvement of AFSA's own communications capability; and AFSA's control and direction of the services.

In summary, while logic seemed to argue for physical merger of the service COMINT activities, the actual establishment of the Armed Forces Security Agency did not occur without major opposition. Most authorities, both military and civilian, opposed its establishment.

With the exception of Defense Secretary Johnson and the initial support of the Army, the majority of the military authorities strongly opposed AFSA as conceptually unsound. The Navy and the Air Force felt that they would lose control of their resources as well as the ability to provide timely tactical support to their field commanders. Of the civilian authorities, the representatives of the Department of State and the CIA proved to be even more strident in their opposition to the establishment of AFSA. They viewed the AFSA concept as detrimental to national intelligence interests, and representative of still another effort by the military to control COMINT resources and intelligence priorities. From their perspective, the AFSA charter completely ignored the roles and authorities of USCIB as established by the NSC in July 1948.

Despite this opposition, Secretary Johnson almost singlehandedly accomplished the establishment of AFSA. Because of the new national interest in unification, as well as presidential support, Johnson pushed through the establishment of the Armed Forces Security Agency on 20 May 1949. In the process of establishing AFSA, Johnson overrode the objections of USCIB, the Navy, and the Air Force. General Canine described the situation best when he characterized the establishment of AFSA "as representing Johnson's shotgun wedding of the Army, Navy, and Air Force organizations."[63]

The original charter, as issued by Johnson, would have permitted a more autonomous role for the new agency. But an almost immediate modification of the AFSA charter by the JCS greatly diminished the authority of AFSA and effectively ruled out any real change of direction for the COMINT structures. Because of the charter change, the Armed Forces Security Council, originally envisaged as an internal advisory mechanism for AFSA, became instead a military tribunal that directed the activities of AFSA – and left AFSA little authority of its own. The military authorities sought to dominate and control all COMINT assets and to prevent them from coming under the direction of the CIA and the Department of State. This proved to be a tactic that they would later regret.

During its three years of existence, AFSA was continually confronted with unresolved operational and jurisdictional problems, many of a critical nature. But AFSA did succeed in accomplishing the physical merger of the COMINT processing activities of the Army and Navy organizations. Organizationally at least, AFSA must be viewed as an important step, no matter how incomplete, in the movement toward the establishment of a national cryptologic effort. The AFSA concept and structure became another building block – and training ground – in the progression toward the centralization of a United States COMINT authority.

Chapter V

AFSA, the CONSIDO Plan, and the Korean War, 1949-1952

From its inception, AFSA faced pressures to restructure it, to weaken its authority, or to abolish it altogether. Almost immediately after AFSA was created, the Army proposed the creation of a new military intelligence agency to be known as the Consolidated Special Information Dissemination Office (CONSIDO). It would control U.S. COMINT requirements and the dissemination of all COMINT. The draft proposal provided for exclusive military control over the U.S. COMINT effort. It quickly drew bitter opposition from the civilian agencies: CIA, the State Department, and the FBI. No sooner had the CONSIDO proposal been rejected than the Korean War broke out and AFSA again found itself in the middle of a major controversy. The war spotlighted AFSA activities and highlighted major weaknesses in the U.S. COMINT structure. Even before the war ended. AFSA became the center of a high-level investigation to reevaluate the role and placement of the U.S. COMINT organization in the overall U.S. intelligence structure.

When AFSA was established in 1949, Secretary Johnson considered a parallel Army proposal to create a Consolidated Special Information Dissemination Office. The Army's plan, with support from the Navy and Air Force, sought to bring all consumers together in a central evaluation unit. The new office would be under military control, and would serve as an "intelligence" counterpart of the COMINT structure. CONSIDO would be charged with the responsibility for performing the requirements and dissemination functions related to the COMINT process for the entire intelligence community. Even more controversial than the original plans for the establishment of AFSA itself, the proposal had far-reaching implications and led to a new struggle between the military and civilian members of USCIB over the control of basic intelligence functions and relationships.[1]

The proposal itself was not new. Near the end of World War II, there had been extensive discussions by the Army and the Navy concerning the concept of a joint evaluation and dissemination center for COMINT product. However, when the services could not reach agreement on a proposal to merge their cryptologic activities, the concept was abandoned. The idea, however, remained alive within each service. Many military officials continued to believe that there should be an integrated COMINT structure charged with responsibility for performing intelligence functions such as the evaluation and dissemination of COMINT product. Three months before the establishment of AFSA, William F. Friedman, chief, Technical Division, played a major role in regenerating the plan. Working in conjunction with the Intelligence Division, Department of the Army, Friedman reworked the proposal, which the Army forwarded to Secretary Johnson a few days before the establishment of AFSA.[2]

The plan recommended the establishment of a new consolidated intelligence agency that would be composed of analysts from the various intelligence agencies, and would operate under the aegis of a military organization – either the director, AFSA, or some other military organization. The chief, CONSIDO, would exercise total control over the development of COMINT requirements, as well as the evaluation, publication, and dissemination of all intelligence based upon COMINT raw material. The proposal envisioned the establishment of a CONSIDO office in Washington and the establishment of overseas branches.[3]

The stated fundamental objective of the CONSIDO consolidation plan was to improve Department of Defense intelligence and contribute to "efficiency and economy." With the establishment of CONSIDO, the COMINT exploitation units of all other departments were to be abolished. CONSIDO was to provide integrated intelligence estimates on all available COMINT and was to reflect the joint view of all intelligence agencies.[4] General Joseph T. McNarney, special assistant to the secretary of defense, became an enthusiastic supporter of the plan, believing it would result in great savings. He directed that the proposal be role of CIA, State, and the FBI in a CONSIDO-type intelligence operation. Pentagon officials recommended that AFSAC, the same committee that directed the activities of AFSA, control CONSIDO. This, of course, would keep control of the organization in the hands of the military authorities. The three civilian elements adamantly opposed exclusive military control over CONSIDO or any organi-

William F. Friedman

coordinated with State, CIA, and the FBI in order to make it as acceptable as possible to them. McNarney, however, showed little real consideration for the civilian views. He remarked that he was "sure that it had been made clear to these agencies previously that the consolidation was a Department of Defense matter and would take place regardless of their opinions."[5]

Secretary Johnson did not immediately endorse the proposal, but instead referred it to the JCS for review.[6] During the next several months, various redrafts emerged throughout the intelligence community for establishing some form of "CONSIDO." The main issues always focused on what organization would control the CONSIDO structure, and the

zation that sought to administer the intelligence requirements process for the total U.S. COMINT effort and that sought to establish strictly military control over policies governing evaluation and dissemination of COMINT information.

The debate finally reached the USCIB, when Colonel James R. Lovell of the Joint Intelligence Committee of the JCS presented the CONSIDO proposal on 2 December 1949.[7] In the ensuing discussions, the USCIB representatives generally reaffirmed their organizational positions. State and CIA indicated they would not support the concept unless they were made jointly responsible with the Department of Defense for running CONSIDO. Both the State Department and the CIA supported the concept of a CONSIDO-type operation, but they opposed the specific proposal because of its military orientation.[8]

W. Park Armstrong, speaking for State, insisted that "the civilian agencies retain their position of equality with regard to their authority and responsibilities in the COMINT field."[9] In a memorandum to the members of USCIB, Admiral Hillenkoetter, DCI, also stated his vehement objections to the CONSIDO proposal. He considered the plan to be in complete derogation of the COMINT roles of the DCI as assigned by the National Security Act of 1947 and USCIB as established by NSCID No. 9. He objected to placing intelligence functions, such as evaluation, correlation, and dissemination of AFSA product, under exclusive military control. Stating that intelligence requirements and priorities were a clear-cut legal responsibility of the CIA, Hillenkoetter further objected to the placement of these functions under the JCS. In short, Hillenkoetter stressed that many of the CONSIDO functions were national in nature and could not arbitrarily be assigned to a structure totally under military control.[10]

Within the new AFSA structure itself existed a wide divergence of opinion concerning CONSIDO. Many of the senior military officials felt that the CONSIDO proposal was conceptually sound. However, Admiral Stone, the director of AFSA, firmly opposed the CONSIDO concept. Because of possible infringements on AFSA's mission and function, Stone argued against the establishment of an additional agency outside the AFSA framework for the production of communications intelligence. Stone took the position that approval of CONSIDO would require a simultaneous revision of AFSA's charter. He stressed that AFSA must be responsive to the needs of the State Department and CIA as well as those of the military.[11]

Because of the objections raised by State and CIA (AFSA was not a voting member of USCIB), USCIB referred the issue to an ad hoc committee under the chairmanship of T. Achilles Polyzoides of the Department of State.[12] The committee continued to struggle for several more months to develop a compromise solution. Although all members of USCIB agreed that the six agencies represented on USCIB might integrate COMINT exploitation activities, they could not agree upon the best form of organization for that purpose. Two quite different proposals finally emerged from the ad hoc committee. One would establish a CONSIDO under the control of USCIB; the other would establish a CONSIDO under the control of the military.[13]

The end of the CONSIDO discussions took place at the fifty-third meeting of USCIB on 14 July 1960. Under USCIB rules, a unanimous vote was required on all USCIB decisions. Since the members were sharply divided on the subordination of CONSIDO, "all the members agreed that CONSIDO should be removed from the agenda, subject to a possible restoration at a later date."[14] This action marked the conclusion of formal discussions over CONSIDO. It never reappeared on the USCIB agenda.

Although CONSIDO was dead, it left permanent scars within the intelligence community. It clearly illustrated the difference of opinion between the civilian agencies and the military establishment about the control of United States intelligence resources.[15] The distinction between "military" and "national" interests began to receive greater attention. The CONSIDO concept soon became the symbol of a new battle to acquire control over the entire intelligence process.

As AFSA struggled to establish itself, the North Koreans launched an attack against South Korea on 25 June 1960. AFSA and the rest of the U.S. intelligence community were caught unprepared. The COMINT requirement in force in June 1950 (issued by the Intelligence Committee of USCIB) stressed primarily the need for information concerning the capabilities and intentions of the Soviet Union and Communist China. South Korea was considered to be outside the defensive perimeter of the United States. The list of countries and subjects "considered to be of greatest concern to U.S. policy or security" included no reference to Korea. Korea was included in the list of secondary requirements. This category was identified as being of "high importance" and as one that required "expeditious" handling to the extent possible. Korea was included as Item 15 in a list of 16 items. Item 15 read: "North Korean-Chinese Communist Relations," and "North Korean-South Korean Relations, including activities of armed units in border area."[16] This low priority statement of interest clearly did not reflect any great consumer interest – nor was it sufficient to justify broad COMINT coverage of North Korean communications prior to the invasion of South Korea.

Because of the absence of consumer intelligence requirements on Korea, AFSA had established no COMINT effort of any kind on North Korean communications.[17] There was no effort on the North Korean problem even on a "caretaker" basis. At the time of the invasion, the only intercept available to AFSA was a limited amount of unidentified traffic. The communications of North Korea first became known to AFSA in 1950 during the course of routine intercept searches for Soviet links. The United States initially intercepted North Korean communications in May 1949 when a search position at an Army installation intercepted an unidentified radio net using Soviet communications procedures. On 21 April 1950, at the request of Army G-2, AFSA assigned an intercept position to the specific mission of searching for and developing information on North Korean communications.[18] As a result of these searches, 220 cipher messages were obtained.[19] For the purpose of identification, this traffic had passed back and forth between the group of AFSA analysts working the Soviet and other problems. By the end of April, the Soviet analysts had concluded that the messages were "probably North Korean," but the two analytic groups could reach no agreement. It was not until after the war had started that the traffic was firmly identified as North Korean.[20]

The outbreak of the war severely strained the limited COMINT resources available to the United States and required considerable diversion of resources and tasking. Not only was the collection and reporting posture on Korean targets very weak in June 1950 but this situation extended to all supporting backup areas as well. There were no traffic analysts working on North Korean communications, no Korean linguists, no dictionaries of the Korean language, no books on the Korean language, no Korean typewriters – and an almost   total absence of knowledge of North Korean terminology.[21]

After the initial attack, AFSA made immediate and drastic adjustments in its COMINT posture, focusing on the now urgent North Korean tasks. Two weeks after the invasion, the intercept coverage of North Korean communications had been increased.[22] Drastic changes in other intercept coverage also took place. For example, COMINT processing activities both in the United States and overseas established twenty-four-hour operations. At the same time, however, there was no compensating reduction in the priorities of other USCIB requirements.[23] As a result, AFSA continued its intercept and reporting of other targets. This coverage included priority reporting on the Soviet Union. The increasing number of priority targets and the limited intercept capabilities presented AFSA with serious problems in allocating its resources for intercept and processing activities. Lacking any unified direction from the consumers, AFSA generally became the tiebreaker in making decisions on conflicting priorities. In other cases it simply deferred to the decision of the military services based upon their intercept capability.[24]

Ironically, the outbreak of the Korean War proved to be beneficial to AFSA in one respect. By bringing national-level attention to the AFSA plight, the war helped to break the budgetary straitjacket that had hampered AFSA. In addition, the possibility that the war might expand to a global conflict led the levying of a multitude of new requirements on AFSA, mainly for intelligence information about the USSR, China, and North Korea. It quickly became evident that the struggling

AFSA needed additional resources. This led to increases in AFSA's authorizations for manpower and facilities, as well as an expansion of the resources for entire U.S. COMINT effort.[25]

The Korean War brought into focus another problem for AFSA that would require more than a simple expenditure of resources to fix. It was organizational and related to AFSA's position in the COMINT structure and its authority to direct the activities of the military services. With the Korean War, AFSA sought to establish itself as the central U.S. authority for COMINT matters. Unfortunately, the conflict between AFSA and the services could not be resolved and greatly impeded AFSA's efforts to fulfill its overall intelligence role. The difficulties stemmed from the inherent weakness of AFSA's charter with its ambivalence about the roles and authorities or the principal participants in the COMINT process.

When Secretary Johnson established AFSA, he designated the new agency as the central defense authority for the communications intelligence activities of the United States, but with one significant exception. The charter excluded from AFSA's control those COMINT facilities and activities that served in direct support of the field commanders for the purpose of providing tactical intelligence.[26] The control and direction of these latter activities remained the responsibility of the military departments. This exclusion clause proved to be highly divisive. It resulted in a continuing and frequently bitter feud between AFSA and the services over who was actually in charge of COMINT. The Army and Air Force, in particular, took advantage of the clause and came into frequent conflict with AFSA over jurisdictional issues. The conflict usually involved tasking matters – questions primarily related to the exercise of operational and technical control over the military field installations. As a practical matter, during the Korean War the Army and Air Force directed their major emphasis toward the development of their own field collection and processing activities, primarily to meet the intelligence needs of their field commanders. In the

judgment of the military COMINT services, they were tasked primarily by military authorities for the intercept of military counterpart traffic at the tactical level. Secondarily, they were tasked by AFSA to intercept other targets of interest to the rest of the intelligence community. This latter category included the intercept of targets identified as being of joint interest, such as civil links. This division of effort resulted in the issuance of separate and independent tasking from both the military intelligence officials and AFSA.[27] The unfortunate split in the exercise of control over the COMINT effort constituted a direct challenge to AFSA's dominant role in U.S. COMINT.

Admiral Stone, alarmed by the continuing feud, sought to clarify and resolve the conflict. Issuing AFSAC 60/26 on 13 September 1950, Stone proposed a more precise definition of the division of responsibility between AFSA and the services, as well as a greater role for AFSA in the tasking of field sites. Stone did not question the need for the services to conduct field processing activities in support of the field commanders, but he maintained that AFSA should be the primary organization to provide the centralized operational direction of field processing efforts. He proposed specific procedures for accomplishing a division of responsibility between AFSA and the services.[28]

In AFSAC 60/26, Stone stated plainly what he considered to be AFSA's role as the centralized COMINT authority in guiding the overall direction of the entire U.S. COMINT effort including those field operations that were delegated to the military departments for direct support purposes.[29] Stone's proposal, at least in the view of the Army and Air Force, was totally unacceptable. It reopened all the earlier arguments over the validity and fundamental purpose of the AFSA concept. The Army's official response, to Stone's amazement, totally rejected the idea of AFSA's exercising any operational direction and control over the Army's field processing effort. Stone noted on his copy of the Army's response that it was the most extraordinary example of a complete reversal of position that he had ever seen. In essence, the Army now claimed that AFSA did not have the responsibility for providing the Army's field commanders with combat COMINT. The Army maintained that it would reserve to itself the right of conducting all of its intelligence operations as it deemed necessary or desirable.[31] Similarly, the Air Force, stressing the need for having its own independent processing capability, rejected AFSAC 60/26, and insisted on controlling Air Force operations for the production of combat air intelligence.[31]

In a reversal of its earlier position, the Navy became the only service to support Stone's paper.[32] This switch was due in large part to Captain Joseph N. Wenger, USN, and his perception of the AFSA role. Wenger's participation in joint Army and Navy discussions extended from the late 1930s to the early 1950s. During the Stone Board deliberations, he provided staff support to the Navy member and was an articulate spokesman against the AFSA concept. Despite his personal feelings about the wisdom of the merger, Wenger, as a deputy director of the new AFSA, played a key role in implementing the merger actions. Motivated mainly by the realities of the situation, however, Wenger came to recognize that even greater centralization actions would be necessary in the future. He became a supporter of the AFSA concept and personally drafted the paper (AFSAC 60/26) that laid out a strong role for AFSA in directing the COMINT effort.[33] This now became the Navy position.

In assessing the Army's strong disagreement with AFSAC 60/26, Wenger remarked about the ironies of the situation. He recalled that the Army had been the main proponent for the establishment of AFSA, whereas the Navy had opposed it, primarily for operational reasons. According to Wenger, the new Army position was completely counter to what was understood to have been the aim of AFSA and to what experience had shown to be the technical realities of COMINT operations. He also saw a real threat to the continued existence of AFSA if the Navy came to support the Army and Air Force position over control of field resources. If this hap-

pened, Wenger speculated that the services would acquire complete independence in large sections of the COMINT problem, thereby depriving AFSA of its primary reason for existence as a military agency – the centralization and coordination of U.S. COMINT.[34] He recognized that the Army and Air Force dissents simply represented a reopening of the earlier controversy over the issues of centralization, but with one radical difference. The Army and the Navy, who had been the principal players during the Stone Board deliberations, were no longer speaking from positions of great operational strength. Each had lost its primary COMINT processing center to AFSA.

According to Wenger, when AFSA absorbed the COMINT processing activities of the Army and Navy, both services lost highly skilled organizations that had taken years to develop and would require many years to replace. Each organization had turned over a major portion of its COMINT-trained manpower, its COMINT machinery, and its COMINT facilities to AFSA in an effort to make that organization work. The Air Force, however, had lost nothing, as it possessed no major resources of its own at the time. In the Navy view, the magnitude of loss for the Army and Navy revolved around the particular needs of each for the production of combat intelligence – which the Navy believed differed greatly for each service. During the Stone Board discussions, the Navy repeatedly stressed this aspect. It asserted that much of the Army's combat intelligence program was targeted against low-level systems, which were exploitable in the field at the tactical level. In contrast, Wenger and other naval officials always maintained that the entire naval problem could be handled properly only in a full-scale technical center, as its complexity required exploitation at the highest analytical level.[35] Based on this rationale, the Navy consistently maintained that it had suffered a greater loss than the Army when AFSA absorbed the two processing centers in 1949.

Navy authorities now perceived AFSA to be a permanent organization and that the Army and

Navy must rely on it for COMINT support. While the Navy authorities had earlier opposed the creation of AFSA, they now felt that it was too late to think in terms of restructuring or abolishing it, particularly during an ongoing war. Consequently, Navy officials, led by Stone and Wenger, supported AFSA and helped it to expand its technical resources and capabilities, including its professional talent and complex machine equipment. They believed that such an expansion of AFSA's technical capabilities would help AFSA meet the future intelligence needs of the three services as well as those of the other members of the intelligence community.[36] In taking this position, Navy authorities made it clear that they would oppose any effort by the Army and Air Force to undo the AFSA merger.

With only the support of his own service, however, Stone made little progress in his efforts to strengthen the concept of a centralized COMINT authority. Discouraged, Stone finally agreed to defer any further consideration of AFSAC 60/26 by the members of the Armed Forces Security Agency Council. Recognizing that he would not win support in the JCS for AFSA control of field relationships, Stone also chose not to submit AFSAC 60/26 to the JCS for decision. Instead he modified his initial broad approach and narrowed his argument to a single issue – Air Force processing of COMINT in the United States.[37]

In a memorandum to the JCS (AFSAC 60/42) of 24 November 1950, Stone took on a head-on challenge from the AFSS concerning the role and authorities of AFSA. AFSS insisted on establishing its own centralized processing activity at Brooks Air Force Base, Texas. Stone asserted that this plan was in direct conflict with the AFSA charter.[38] Although the JCS never took official action, the AFSS canceled its plan to develop Brooks as a centralized processing center. The Air Force, however, soon announced new plans (AFSAC 60/49) to establish additional major processing centers.[39] This action strongly reinforced the concept that the services were in charge of conducting tactical field operations.

With this sequence of actions involving AFSAC 60/26 and AFSAC 60/49, the Army and the Air Force prevailed over AFSA in their insistence that the services, not AFSA, should maintain the dominant role in controlling the activities of their field resources. The distance from Washington, as well as the need for timeliness in reporting the intelligence information, contributed to this victory. Moreover, this was an era when service collectors were tasked primarily with the intercept of the communications of their foreign counterpart service, and secondarily with the intercept of joint targets (e.g., civil). Thus, for a combination of reasons, the military commanders in Korea would rely primarily on their respective military services for tactical intelligence rather than on AFSA. While AFSA officials continued to believe that AFSA should exercise operational and technical control over all COMINT activities, they did not have sufficient power and authority to impose their will on military COMINT services.

As the Korean War continued, both the Army and the Air Force organizations expanded their field intercept capability and established their own field processing activities. Despite the continuing controversy, the operational elements of AFSA and the military services worked together in great harmony. In their day-to-day coordination on operational matters, they demonstrated a strong spirit of mutual cooperation and assistance. They freely exchanged information and technical details related to the collection and analytic processes, including such items as translations, cryptanalytic recoveries, and intercept data. There were also personnel exchanges between AFSA and the services over a broad range of operational functions. For example, AFSA sponsored continuing programs for AFSA personnel to perform temporary duty assignments in the field in order to assist the services – and for military personnel to participate in orientation and training programs at AFSA prior to their assignment to the field. By the time the war ended, these field programs became major focal points for COMINT reporting and provided unique intelligence contributions to the field commanders.

From the beginning of the war, it was evident that U.S. combat forces in Korea would rely on the individual service processing units in the Far East for tactical COMINT support. Starting with only one intercept station, the Army COMINT organization ultimately acquired the largest contingent of field units in support of U.S. operations in Korea. The Army set up its field headquarters, the Army Security Agency, Pacific (ASAPAC). ASAPAC also served as a major processing center in the theater, directing Army fixed intercept sites located in Hawaii and Korea. In addition, ASAPAC established a number of advance detachments in Korea. By 1951 the 501st Communications Reconnaissance Group headed all Army COMINT units in Korea. Subordinate units were designated Communications Reconnaissance Battalions or Companies (CRB, CRC). ASAPAC exercised overall control of Army COMINT operations in Korea, including South Korean detachment.

At the outbreak of war, there was only one Air Force unit in the Far East, with scattered detachments. After the North Korean attack, the Air Force established a detachment in Korea for intercept and reporting and for direction of a South Korean unit. In 1951 the Air Force began to deploy smaller teams to Korea for the production of tactical COMINT for the 5th Air Force. Eventually, the Air Force established an area headquarters which gradually assumed responsibility from ASAPAC for the intercept and processing of the ground communications of the North Korean Air Force.

The reliance on counterpart coverage, coupled with the small number of North Korean or Chinese naval forces involved in the war, precluded any major role for U.S. naval COMINT units. In the course of the war, however, a Navy radio facility did provide important assistance to the overall Far East COMINT effort. Detachments afloat also contributed unique intercept of Soviet nets and Chinese activities.

At the Washington level, AFSA also attempted to improve its relations with consumers. Stone, in his role as AFSA director and executive agent of USCIB in COMINT matters, attempted to keep the consumer community current on all AFSA actions. He not only encouraged the establishment of consumer liaison offices at Arlington Hall but also promoted an expansion of the direct dialogue between intelligence analysts and COMINT producers. Traditionally, the two performed their tasks with little interaction. This new dialogue took place in an era when the COMINT community was particularly sensitive to the release of "technical" information or "tech data" to consumers. The term "tech data" generally referred to the operational details of the intercept, analysis, and translation process. It included specifics related to collection sources, callsigns, identification of communications links, identification of cryptographic systems, and other details of the analytic process.[39] The issue over providing technical data to the consumers was never resolved. It continued to be a source of conflict between COMINT producers and consumers. Many consumers, especially CIA, demanded the data. The producers opposed providing it.

Initially, Stone's push for a freer interchange between producers and consumers caused considerable opposition within the COMINT family itself. Many AFSA personnel viewed the move toward closer dialogue as blurring the distinction that traditionally existed between the roles of the COMINT analyst and the intelligence analyst. The COMINT analyst was to provide only raw data. The intelligence analyst produced finished intelligence. Because of the war and its pressing priorities, however, such distinctions soon dissipated.[40]

Despite such problems, there existed major areas of cooperation among the intelligence producing agencies in Washington. Long before the establishment of AFSA, the Army G-2 and the Office of Naval Intelligence established collocated offices with their service counterparts at Arlington Hall Station and the Communications Supplementary Activity, Washington. Within each service, the consumer and producer elements developed harmonious working relationships and operated with a minimum of correspondence or formality. These consumer contacts extended not only to the COMINT processing elements but to the policy-making officers or the COMINT organizations as well.[41]

When AFSA was established, many of the Army and the Navy COMINT officials simply moved over to new positions of authority corresponding to their previous roles in their old agencies. In this manner, the liaison arrangements between the military producers and consumers operated smoothly during the transition period.

This liaison arrangement worked particularly well for the Army's Special Research Branch (SRB) during the Korean War. One of the most active consumers was G-2. Early in the war, SRB collocated personnel in AFSA's North Korean section, where they worked closely with the AFSA COMINT analysts in almost all phases of the exploitation process. Since the SRB representatives scanned all translations and traffic analysis reports prior to publication, they were able to develop a unique perspective of COMINT operations. In addition to this direct participation in COMINT activities, SRB served as a general channel for the dissemination of COMINT from AFSA to the Far East commands.[42]

In contrast to the long-standing working arrangements of the Army and Navy consumers with their military COMINT counterparts, the Air Force started from scratch. When the Air Force became a separate service in 1947, the Army continued to provide intelligence support to the new service on an interim basis. Within a year, the Air Force activated the AFSS, its own COMINT processing organization that began operations at Arlington Hall Station on 1 February 1949 and relocated to Brooks Air Force Base in Texas in May 1949.[43]

Responding to the demand for more timely COMINT product, the Air Force relocated a part of its Office of Intelligence (AFOIN-C/R), under Colonel Horace D. Neely, to Arlington Hall in 1950.

The purpose of the move was to enable Air Force intelligence analysts to work more closely with AFSA, as well as with the Army (Special Research Branch) and Navy (OP-922YI) intelligence organizations. The operations of the new AFOIN-C/R soon paralleled that of other consumers, with its intelligence analysts consulting freely with processing personnel in AFSA's working areas.[44]

In time, however, the AFSS assumed administrative control of all Air Force COMINT activities in the Washington area. By February 1952 all of the Air Force units were combined into the Washington, D.C., Control-Collection Office (WDC/CCO) of AFSS headed by Colonel James L. Weeks. This organization represented the Air Force in a dual capacity, both as a producer and user of COMINT. As a COMINT producer, it worked with AFSA's Office of Operations in the development of intercept assignments and served in a general liaison capacity as AFSA's point of contact with the AFOIN-C/R and the AFSS. As a consumer, it produced finished intelligence and represented the Air Force on USCIB and AFSA boards and committees. By 1952 the Air Force was the largest consumer delegation resident at AFSA, with well over 100 people assigned to its control-collection office.[45]

Although the Department of State's active involvement in postwar COMINT dated back at least to 1945, when it joined with the Army and Navy organizations to form STANCIB, the establishment of a full-time State liaison officer at Arlington Hall did not come about until the Korean War.[46] It grew out of discussions of the USCIB on 14 July 1950 concerning mobilization and later discussions on the subject between Rear Admiral Stone and W. Park Armstrong, the State Department's representative on USCIB. When Armstrong suggested State's willingness to provide direct financial support to AFSA for increasing COMINT output, Admiral Stone countered by suggesting that the establishment of a State Department liaison group in AFSA might be more useful, Stone felt that State could assure fulfillment of its intelligence requirements by an "on-the-spot audit of the COMINT production program," and could also handle such working problems as determination of priorities and transmission of collateral.[47]

Armstrong accepted Stone's offer, but the State Department took no official action until a year later. On 15 July 1951, the department established a liaison unit at Arlington Hall consisting of John Crimmins and three assistants. The unit had no specific title at first, but was later designated the "Field Branch, Special Project Staff." The new Field Branch followed the pattern of operation of other liaison offices, but on a much smaller scale. It worked directly with AFSA 02, advising it of State Department COMINT requirements. It also worked directly with the analytic and collection elements of the AFSA Office of Operations in order to provide more detailed information about the requirements transmitted through the USCIB committee structure.[48]

AFSA's relationship with CIA underwent a greater change than with any other consumer group. Although initially uncertain of its charter and authorities, the new CIA gradually began to seek greater participation in COMINT activities. In response to a request from DCI Admiral Hillenkoetter, in 1948 USCIB authorized CIA direct access to COMINT activities. USCIB also authorized, under certain circumstances, the direct participation by consumers in actual COMINT production activities. It represented a major change in the consumer-producer relationship.[49]

Shortly after the physical merger of Army and Navy COMINT processing activities in 1949, CIA, with the concurrence of Admiral Stone, moved into its first official liaison office at Arlington Hall. Starting in late 1949, John S. Ward served part-time as the first CIA liaison officer. During this period, the main CIA interest was Soviet plain text. Based upon the earlier USCIB decision, CIA arranged with AFSA for the assignment of some CIA personnel to the Russian Language Branch (AFSA 246) where they were integrated within the

AFSA structure. With the support and guidance of AFSA personnel, CIA carried on its own research in the plain text unit. In addition, CIA assigned a limited number of personnel to work in the collateral and COMINT files of the Central Records Office (AFSA 25).[50] With the onset of the Korean War, CIA merely expanded its cooperative effort with AFSA.

By 1951 the entire consumer community had liaison offices at Arlington Hall. Because of their well-established relationships with their military counterparts in COMINT, however, and because of AFSA's military orientation, the Army and the Navy representatives enjoyed a greater access to AFSA's operational offices than those from the Department of State and the CIA. AFSA, as a predominantly military organization, tended to be more responsive to the military when determining intercept priorities. Nevertheless, the representatives of the civilian agencies welcomed the establishment of a beachhead at Arlington Hall and the opportunity for direct and continuing dialogue between the producers and consumers on intelligence matters. While these measures contributed to a less confrontational attitude within the intelligence community, representatives of the civilian agencies still felt uncomfortable with the basic design of the U.S. COMINT structure and their lack of influence with AFSA. It was a continuation of the strong objections expressed by officials of the State Department and CIA when AFSA was first established as an autonomous intelligence arm of the Department of Defense.[51]

Despite problems, AFSA did succeed in making significant changes in its relationships with the consumers. Stone designed a new "open-door" policy for COMINT relationships with consumers, both military and civilian. He encouraged and facilitated the establishment of consumer offices at Arlington Hall in an effort to improve the dialogue between the producers and consumers. Both seemed to benefit. The consumers, by virtue of their physical presence at the COMINT center and by their participation in AFSA's priority mechanism, saw at first hand the inner workings of the AFSA structure – as well as the major problems confronting AFSA. In addition, the consumers began to display a greater appreciation of COMINT as a unique and valuable source of intelligence information. Although the concept of exercising "oversight" over the COMINT structure had not yet materialized, the intelligence community began to move in the direction of discussing more critically and more openly the quality, the utility, and the timeliness of COMINT reporting. Joint community actions and discussions of an evaluative nature now occurred more frequently. For example, the USCIB Intelligence Committee began publishing a monthly report listing the total number of COMINT messages published. It was broken down by country and series and indicated whether the messages were plain text or encrypted. The report also showed the statistical improvements in volume over the previous month. Despite the improving relationships, AFSA made no real progress in resolving the serious management and operational problems affecting its relationships with the military COMINT services by the summer of 1951.

The basic question of operational control of service intercept facilities remained unresolved. Stone, the first director of AFSA, in fact controlled fewer intercept positions than his predecessor, the Coordinator of Joint Operations under the earlier Joint Army-Navy Operating Plan. Under the Joint Operating Plan, the CJO had direct access to joint intercept positions and a restricted access to all other intercept positions. But Stone, as AFSA director, did not have this same access. As noted earlier, a large number of the intercept facilities, namely the mobile collection sites, were removed totally from the control of AFSA. These sites were under the exclusive operational control of the services. Even in exercising AFSA's authorized control over the fixed intercept sites, Stone had to operate under a rigid set of arrangements and rules prescribed by each service.[52] The system was not designed to enhance timely reporting of COMINT information.

Another nagging problem for AFSA was the extent to which the services conducted their own autonomous processing activities in the field. While AFSA's protestations over the control of the field processing centers had abated somewhat with the Korean War, the problem remained. The services continued to dilute AFSA's role as a central authority. In particular, problems with the Air Force intensified as the AFSS persisted in its efforts to acquire primary control over the total air problem. Even more damaging to U.S. COMINT than the friction generated by this issue was the broader question of duplication and waste of resources both in the field and stateside processing centers.

Benson K. Buffham   Herbert L. Conley

For example, in early 1951 ASAPAC and AFSS both covered Chinese Communist and Soviet targets. Major General Charles A. Willoughby, G-2 of the General Headquarters, Far East Command (FECOM) requested that a high-level AFSA team visit the theater to assist G-2, FECOM, and ASAPAC in a consultative capacity. On 12 March 1961, Stone sent two of his senior officials to the Far East Command to brief Willoughby and review the field operational problems. Benson K. Buffham, assistant chief, General Processing Division, and Herbert L. Conley, a senior manager and collection specialist, went to Korea. In their final report of 2 April 1951, Buffham and Conley cited the duplicative efforts of AFSA, ASAPAC, and the AFSS on the North Korean problem. Among their recommendations they proposed that ASAPAC and AFSS divide and coordinate their efforts, particularly on certain

Soviet and Chinese Communist problems. Despite the urgency of their recommendations, duplication continued until March 1952, when the AFSS assumed total responsibility for the Chinese Communist and Soviet problems.[53]

During 1951 AFSA confronted a number of operational problems as well. U.S. COMINT contributions to the war effort were far below the achievements of COMINT during World War II.[54] Suffering from a shortage of intercept facilities, short tours of duty by military personnel, and difficulty obtaining linguists, AFSA could not fully exploit COMINT possibilities during the war.[55] Moreover, by the fall of 1951, the North Koreans introduced new and more sophisticated cryptographic systems.

As the war dragged on, AFSA concentrated its efforts at increasing the flow of tactical COMINT from the services to the field commanders. The real COMINT success story of the Korean War proved to be in the area of tactical support. Because of the nature of the land war, coupled with the assignment of counterpart coverage to each service, the Army and Air Force controlled, almost totally, the intercept coverage and reporting on the Korean and Chinese targets. In providing direct support to the Eighth Army and the UN Forces, the Army and Air Force are generally acknowledged as having made the principal military contributions to the COMINT effort in Korea. An unpublished NSA review of COMINT in the Korean War written in 1953 emphasized this point:

> *Perhaps the most interesting development of COMINT in a tactical support role was the sucessful expansion and utilization of low-level voice intercept. In August 1961, the effort was of the most rudimentary nature—but the nature of the intelligence provided was of such immediate tactical value to the corps, division and regiment commanders that those commanders clamored for additional support. By the end of October 1951, seven low-level voice teams had been formed in support of the U.S. I and IX Corps. By June 1952, there were ten teams in action along the Eighth Army front.*
>
> *These teams were able to advise frontline unit commanders of imminent enemy artillery or infantry action and their advance warnings through the balance of the war were instrumental in the success of UN counter actions and the saving of many UN lives.*[56]

The results of the COMINT successes in the exploitation of Korean and Chinese military and air force communications, to most intelligence consumers, still looked extremely thin. Many officials in the U.S. intelligence community, aware of the impressive contributions made by COMINT in World War II, complained about AFSA.

While the military services were, in general, satisfied with AFSA's attention to their intelligence requirements, State Department and CIA officials were not. They felt that their intelligence requirements were not being met. They complained about their lack of input in establishing intelligence requirements and AFSA's lack of authority in translating these priorities to the collectors. AFSA officials made vague attempts to pacify the civilian complaints stating that "we will take care of it," that "of course, this is of interest to all of us," and "you can be sure that you will get your share." But this did not satisfy the State Department nor the CIA. In truth, the civilian agencies were correct. There was no existing mechanism whereby the rapidly growing CIA could express its needs for information to the cryptologic community. Similarly, for the State Department, there existed no regularized channels it could use with any degree of assurance needs would even be considered by AFSA.[57]

By 1951 the flaws of the AFSA experiment were clear. The division of responsibility between AFSA and the services prevented AFSA from undertaking any serious new initiatives to improve the total U.S.

COMINT product. As a direct corollary of this, Stone's lack of authority over the services greatly diminished the quality and timeliness of COMINT reporting and resulted in duplication of COMINT coverage. Fractionalization prevented AFSA from operating as a centralized COMINT organization.

Because AFSAC invariably supported the service viewpoints rather than AFSA's on the issue of AFSA's authority, AFSAC was of no assistance to AFSA in resolving the serious jurisdictional disputes. The last avenue of appeal was to USCIB. However, USCIB was little more than a coordinating body with no real authority over the AFSA structure itself or its organizational role. Although USCIB had a vital interest in the intelligence produced by the military components, it could not resolve the jurisdictional issues between AFSA and the services.[58]

In summary, AFSA received persistent criticism from the time of its creation. The U.S. Army's CONSIDO proposal was an attempt by the U.S. military to establish a separate COMINT intelligence agency parallel to AFSA that would maintain strict military control over most U.S. COMINT sources. It met with bitter opposition from the civilian intelligence agencies such as CIA and the Department of State. Although the proposal was defeated, the deliberations concerning CONSIDO reinforced the generally hostile climate existing in intelligence matters, and the continuing concerns of CIA and State that their intelligence needs were not being met.

The Korean War proved to be a major turning point in the history of the U.S. COMINT structure. At the outbreak of the war, glaring weaknesses appeared in the AFSA structure. There were major problems with resources, intercept and reporting capabilities, and the cryptanalytic attack itself. Most importantly, the war illustrated AFSA's inability control the COMINT organizations of the services and its inability to control and direct U.S. COMINT resources in an efficient, effective manner. Despite AFSA's attempts at increased coordination and some success with the exploitation of low-level technical communications, U.S. policy makers came to see AFSA as a basic failure. It did not or could not duplicate the COMINT successes of World War II. It thus became the centerpiece in a high-level investigation to reform and redirect the entire U.S. COMINT structure.

Chapter VI

The Brownell Committee and the Establishment of NSA, 4 November 1952

On 24 October 1952, President Truman issued an extraordinary directive that changed the organization and direction of the U.S. communications intelligence structure and laid the policy framework for the modern system. Truman stated that the communications intelligence function was a national responsibility rather than one of purely military orientation. This triggered actions that reorganized the U.S. military COMINT effort and strengthened the COMINT roles of the USCIB and the NSC and brought a wider role for the civilian agencies in U.S. COMINT operations. The president's memorandum also contained the first reference to a "National Security Agency," to be established in place of the Armed Forces Security Agency. Under Truman's directive, the Department of Defense became the executive agent of the government for the production of communications intelligence information, thereby removing the JCS as the controlling authority for the COMINT process.

Truman's directive stemmed from the recommendations of a presidential commission known as the Brownell Committee. Truman established the committee to conduct an investigation of the efficiency and organization of the entire U.S. communications intelligence effort. By December 1951 AFSA's disappointing wartime performance had been brought to the attention of the White House, and Truman responded by calling for a complete review of the COMINT structure. Setting up a mostly civilian committee, however, caused great alarm within the military, particularly in the JCS. In February 1952 the JCS complained that it had no part in the deliberations leading to the committee's establishment and that the U.S. military had been excluded from membership on the committee and its support staff.

The final Brownell Report emphasized the need for the establishment of one organization to manage the communications intelligence activities of the government. The report provided a strong indictment of service unification as it existed under AFSA as well as an indictment of the management and policy echelons existing above AFSA. The report recommended a complete reorganization of the U.S. COMINT effort, and provided a blueprint for the new structure. As its main theme, the Brownell Committee pressed for the elevation of the COMINT structure to a new status, requiring national-level attention and interest. It also spoke out against the almost total autonomy of the military in COMINT matters. This chapter details the history of the creation of the Brownell Committee, its report to the president, and subsequent acts that had a major impact on the U.S. intelligence community and led to the creation of the National Security Agency.

By 1951 the CIA and the Department of State representatives of USCIB felt vindicated in their original opposition to the establishment of AFSA. The problems associated with the operations of AFSA had grown considerably and now extended to a broad range of intelligence community relationships. Organizationally, the fundamental issue over the division of responsibility between AFSA and the military COMINT services appeared to be no closer to a solution. The Korean War evoked new criticisms of the AFSA structure. A spirit of disunity and turmoil characterized the activities of the entire intelligence community.

The major players of the intelligence community were locked in a struggle over "who was in charge" and over the acquisition of expanded responsibilities and authorities. The military and civilian agencies continued to argue over basic jurisdictional and organizational relationships. These disputes greatly complicated the entire intelligence picture. The new CIA, seeking to expand its role, pushed for greater authority in the total intelligence process. While the Department of State had different intelligence interests than CIA, it generally aligned itself with CIA on most issues during the USCIB meetings and in the protests over the lack of civilian/military equality in the COMINT field.[1]

In particular, the vigorous and heated discussions over the establishment of AFSA, and later over the JCS proposal to establish a new military intelligence agency, CONSIDO, constituted head-on challenges, not only to the role of CIA in the intelligence field, but to the authority of the National Security Council as well. Although the CONSIDO proposal was dropped, CIA and State viewed it as an attempt to acquire a dominant and proprietary role for the military in such intelligence functions as estimates, evaluations, and dissemination of intelligence.[2] CIA and State perceived a constant erosion in their ability to get the COMINT structure to consider and satisfy their intelligence needs.[3]

The stage was set for reform. But what shape should the reform take? What were the avenues for resolution of the "AFSA problems" – and the increasing tension within the community over the ownership and control of ancillary COMINT functions? AFSA by itself could not resolve the many managerial and operational conflicts. Nor did there appear to be any likelihood of a solution emanating from USCIB, which remained powerless because of its limited charter and military-dominated membership. Because of its membership majority, the military organizations were able to control the board in its working-level committees.[4]

Given the rigidity of the existing COMINT structure, CIA and State officials probably concluded that further dialogue would be fruitless. Taking direct action, they pressed for fundamental changes in the intelligence structure. It was an opportune time. The Truman administration was extremely budget conscious and was known to favor centralizing and consolidating intelligence responsibilities and functions.[5]

There was also a new DCI on the scene. General Walter Bedell Smith, USA, appointed by President Truman on 21 August 1950, succeeded Admiral Hillenkoetter as the fourth director of Central Intelligence. The appointment of Smith represented a significant change of leadership for CIA and foretold a change in the CIA posture as well as in its approach to intelligence community relationships. Having served as chief of staff under General Dwight D. Eisenhower in the European Theater from 1942 to 1945, and as U.S. ambassador to the Soviet Union from 1946 to 1949, Smith had gained the personal support and confidence of President Truman. Known to be an exceptionally strong and forceful executive, Smith was highly respected by Eisenhower and other top military and government officials, particularly for his organizational talents.

General Walter Bedell Smith, DCI

During his three-year tour as DCI, Smith played a preeminent and rigorous role in the organizational evolution of the DCI and CIA roles. This, in turn, had a major impact on the entire intelligence community. Aggressive action became a keystone of Smith's new approach.[6]

With a view to reforming the COMINT structure, Smith brought the AFSA problem to the atten tion of the NSC on 10 December 1951 and recommended an overall survey of the COMINT structure. His recommendation for such a survey was based on a study by Kingman Douglass, who was then the CIA COMINT officer. The NSC, in turn, forwarded Smith's proposal to President Truman. Three days later, Truman directed Secretary of State Dean Acheson and Secretary of Defense Robert Lovett, assisted by Director of Central Intelligence Smith, to review in depth the communications intelligence activities of the United States government.[7]

Acheson, Lovett, and Smith responded by creating a high-level committee to accomplish the survey. On 28 December 1951, they created the Brownell Committee, headed by George A. Brownell, a prominent New York City attorney. Brownell was assisted by Charles E. ("Chip") Bohlen, counselor, State Department; William H. Jackson, special assistant to the DCI; and Brigadier General John Magruder, USA (Ret.), special assistant to the secretary of defense. The CIA and the Department of State provided the staff members of the committee: Lloyd N. Cutler and Harmon Duncombe, CIA; Grant C. Manson and Benjamin R. Shute, State. All of the staff members had served previously in the Special Intelligence Branches of the Army or the Navy. The Brownell Committee and its support staff took up residence at CIA and were administratively supported by CIA. The carefully tailored composition of the Brownell Committee and its supporting staff omitted one important group. The military authorities, who heretofore had dominated the U.S. COMINT structure, were not included at any level in the actual review process.[8]

Acheson, Lovett, and Smith directed the Brownell Committee to undertake a survey of the COMINT structure and to submit recommendations on two general subjects:

(1) the needs of each governmental department and agency for the production of departmental intelligence, and of *the Director of Central Intelligence for the production of national intelligence, and*

(2) the most effective allocation of responsibilities for COMINT activities, and the extent to which these activities should be performed by a single department or agency as a service of common concern - and to which department or agency such assignment should be made.[9]

In short, the committee was to look at centralization and placement of the entire U.S. COMINT effort in the U.S. intelligence community.

George A. Brownell

Meeting of the National Security Council, January 1951

Charles E. ("Chip")
Bohlen

William H. Jackson

BGen John
Magruder, USA,
Ret.

The establishment of the Brownell Committee provoked immediate outcries within the U.S. military. Four weeks after the creation of the committee, the Joint Chiefs summoned the director, AFSA, the three service directors of intelligence, and the deputy director, Joint Intelligence Group, JCS, to a special meeting to discuss the Brownell Committee's investigation of COMINT activities.[10] On 4 February 1952, Major General Ralph J. Canine, USA, Rear Admiral Frank L. Johnson, USN, Major General Alexander R. Bolling, USA, Major General John A. Samford, USAF, and Brigadier General Richard C, Partridge, USA, met with the Joint Chiefs. The message of the meeting was clear – the Joint Chiefs were alarmed over the activities of the Brownell Committee. The service chiefs complained that they had not been consulted about the investigation prior to its conception, that they had no representation on the board, and that the line of questioning indicated a possibility that the board would recommend transfer of AFSA from the control of the JCS. Despite their forebodings, however, the options open to the JCS remained limited, particularly in light of the fact that the committee was already in operation. Without the unanimous support of the NSC or the USCIB, the JCS was obviously in no position to risk a head-on challenge with the president or the secretary of defense.[11]

In trying to salvage something, the JCS had little choice but to settle for some rather pro forma actions. They asked General Omar Bradley, chairman, JCS, to meet with Secretary Lovett and "again express the considerable concern of the JCS over the possible transfer of AFSA from their jurisdiction." They also decided that the JCS would make its own full-scale review of the AFSA problem – but at a later date – to determine whether more authority should be given to the director "so he could actually control in one organization the COMINT effort of the United States."[12] While this projected study was a tacit admission that all was not well within the COMINT structure, it reflected a glimmer of JCS optimism that somehow the Brownell effort would wither and die without producing any tangible results. The JCS never got around to conducting its projected study. The Brownell Committee moved too swiftly.

The final Brownell Report, submitted to Acheson and Lovett on 13 June 1952, confirmed that JCS apprehensions about the loss of AFSA were well founded. The report completely demolished the concept of unification of the three services as it existed under AFSA. The committee concluded that the structure of COMINT activities did not reflect unification under single control, but rather a structure of four associated agencies – one of which, AFSA, performed limited functions in ways acceptable to those who controlled the other three. In short, it was a military organization controlled by the military.[13]

The report hammered out the theme that the director, AFSA, had insufficient authority or control over the COMINT activities of the three services. It noted, ". . . that for all practical purposes the AFSA charter made AFSAC (which is nothing except a committee made up of the three services) the boss of AFSA, which in turn is completely dependent upon the Service organizations for all its communications and practically all of its collection of COMINT."[14] In reviewing the management framework in which AFSA operated, the report noted that the director had to spend much of his time and energy on cajolery, negotiation, and compromise in an atmosphere of bitter interservice rivalry. According to the report, the director of AFSA had no real degree of control over the service COMINT units – but rather was under their control by virtue of their representation on AFSAC. His only appeal was to the same three services.

The committee also had harsh words for other parts of the U.S. intelligence structure. It strongly criticized the management and policy structures existing above AFSA (USCIB, JCS, and AFSAC) for their total lack of effectiveness in providing guidance, direction, and management support to AFSA. Noting that "the U.S. Communications Intelligence Board (on which the State Department, the Central Intelligence Agency, and the Federal Bureau of Investigation, as well as the three services, are represented) has inadequate authority and has become an ineffective organization," the committee concluded that "the COMINT effort of today has too many of the aspects of a loose combination of the previous military organizations and too few of a true unification of the COMINT activities and interests of all the interested departments and agencies."[15]

In reaching its conclusions, the Brownell Committee stated that its basic thinking was influenced by "two controlling but somewhat conflicting factors."[16] As a first and fundamental premise, it believed that all of the interested services and agencies should have a voice in determining AFSA policies and giving it guidance. Second, as a counterbalance to this, and in order to strengthen the COMINT structure, the committee stressed that AFSA should be placed under a single governmental department for administrative purposes. This line of reasoning signaled, for the first time, the identification of COMINT resources as being

national in nature. It signaled too the probable end of the era of exclusive military control of COMINT resources. Thus, in the view of the committee, the removal of the COMINT structure from JCS control was a necessity. Ideally, and in order to provide an effective COMINT response to the intelligence requirements of all consumers, the responsibility for the COMINT function should be centralized in a neutral governmental agency that would have some latitude in its operation.

The Brownell Report concluded that a point had been reached in the evolution of the United States intelligence community that now made it essential "to carry further the 1949 reorganization of the COMINT structure."[17] It proceeded to outline many actions that should be taken "to strengthen AFSA itself and to increase its authority over COMINT results." These recommendations, directed essentially toward a reorganization and unification of the COMINT structure, addressed a number of broad COMINT relationships, both within and outside the COMINT structure. The scope of the recommendations extended into the major operational and management phases of the COMINT business, including the production of COMINT, the centralization of COMINT authority, the management of COMINT resources, and policy oversight of COMINT resources.[18]

From the outset, the Brownell Committee recognized that complete unification would be impossible because of the dependence upon the military structures to man field stations. Consequently, Brownell concluded that the service units must retain their own authorities and responsibilities within their military departments. To ameliorate this and to assist AFSA in its mission of providing effective, unified organization and control of COMINT, the Brownell Report recommended that AFSA should have operational and technical control over all the COMINT collection and production resources of the military services. The Brownell Committee also supported the services' traditional position that they must control the close and direct intelligence support of the forces in the field. The committee fully recognized that it was creating a problem area between the new central authority and the services, but concluded that a solution could be found "with sincere and intelligent cooperation between the commanders involved."

Specifically, the committee proposed structural changes affecting the three levels of AFSA's organizational relationships: below AFSA (i.e., external service relationships); within AFSA; and above AFSA (i.e., USCIB and the Department of Defense). As to the first, the committee recommended that AFSA be established as the keystone of the COMINT organization – with its mission clearly defined by presidential memorandum. As outlined by the committee, the mission statement would give AFSA the responsibility and authority for providing a unified organization and control of the COMINT activities of most of the federal government.[19]

For those changes projected within AFSA itself, the committee concentrated principally on "people" considerations. It recommended that the director be a career military officer of at least three-star rank, with a tour of at least four years, rather than the two-year rotational tour established by the AFSA charter. The option of appointing a civilian director was left open if the particular circumstances warranted. The military director would have a career civilian as deputy – with the converse to apply in the event of a civilian director. In stressing the need for the development of a strong personnel program, the final report included a major discussion of a broad range of personnel considerations. The report concluded that the existence of a well-rounded personnel development program was essential to the future growth and success of AFSA. It stressed the dimensions of the "people" problems then existing at AFSA, including the exceedingly high rate of turnover among AFSA civilians and the lack of professional and managerial opportunities for the civilian workforce. The report strongly recommended that AFSA initiate greatly expanded efforts to develop career and profession-alization programs for civilian and military personnel – at both the managerial and professional levels.[20]

In discussing projected organizational changes "above the AFSA level," the committee remarked that this category represented the single most difficult question it confronted. As the cornerstone of its recommendation in this area, however, the committee expressed no qualifications about the need for severing the relationship between the JCS and AFSA. It recommended the immediate termination of the 1949 "experiment" that had placed AFSA under the control of the JCS. As a corollary, the committee proposed the abolition of AFSAC. In place of JCS control, the Brownell Committee suggested that AFSA be directly subordinated to the Department of Defense as the executive agent of the government for COMINT activities.[21]

In much the same vein, the report recommended a revitalization and restructuring or USCIB. The report proposed sweeping membership changes including a significant decrease or military representation as well as the simultaneous elevation of AFSA to a position of full voting membership on the new USCIB. The report recommended that USCIB's membership consist or a representative of the secretary of defense, the secretary of state, the DCI, the chairman of the Joint Intelligence Committee, the FBI, and the director of AFSA. It further recommended that the DCI become the permanent chairman of USCIB. This would bring to an end the long-standing practice of selecting a new USCIB chairman annually, based upon a vote of the membership. The report also included a new procedural methodology to govern USCIB operations – with the objective or giving it a greater responsibility "for policy and coordination" in COMINT matters. It also proposed to establish a majority-rule principle when voting in USCIB on matters under the jurisdiction of AFSA. For those COMINT matters outside the jurisdiction of AFSA, however, it proposed retention of the rule requiring unanimous agreement of the members.[22]

Finally, in a related matter, the committee discussed the existing manpower levels and dollar expenditures of the entire U.S. COMINT effort in order to provide an indication of the cost to the government for the production of COMINT. It developed the data based upon figures received from AFSA and the three services. While acknowledging that its figures were possibly only little better than informed guesses, the committee cited these as representing a reasonable approximation of the resources spent in acquiring and processing communications intelligence information.[23]

In this investigation, the committee established the combined manpower levels for AFSA and the three cryptologic services, and it estimated direct cost expenditures for the combined activities. Because of security factors, the report noted that the cryptologic budgets were not subject to the usual checks and balances normally associated with the budget cycle processs. Considering this as well as the magnitude of the expenditures for COMINT, the committee concluded that the fiscal process represented another compelling reason for establishing a strong and responsible AFSA – operating under the positive guidance of a policy board acting with real authority.[24]

On 23 June Secretary Lovett sent the Brownell Report to General Canine for his personal comments and recommendations. In a lengthy response, Canine enthusiastically supported the major conclusions and recommendations of the report. Canine did, however, take issue with some aspects of the report. The committee had proposed an extension of the responsibilities of AFSA to include a communications security responsibility for the entire United States government rather than merely the Department of Defense, as was then the case. Canine agreed in principle with the proposal but pointed out that the committee had included very little on communications security in its report. He recommended that the proposed directive be confined to COMINT, with communications security to be made the subject of another study and to be addressed in a separate directive. Canine also

Major General Ralph J. Canine

argued strongly against retention of the rule of unanimity – even for COMINT matters outside the jurisdiction of AFSA. He pointed out that this would only serve to perpetuate one of the chief difficulties that had hampered USCIB in the past. He urged the acceptance of a majority-rule principle to govern all USCIB decisions. Lastly, Canine pointed out that the Brownell Report had omitted the generally accepted identification of COMINT activities as being outside the framework of security rules governing other intelligence activities. Canine recommended that the same stringent security considerations then existing in NSCID No.9 be carried over into the new directive.[25]

During the next four months, extended negotiations took place among the representatives of the Department of State, the CIA, the Office of the Secretary of Defense, and AFSA over the exact wording of the implementing presidential memorandum. Some of the issues discussed included the definitions of communications intelligence and finished intelligence; principles to govern the production of COMINT by the cryptologic structures; and a number of other policy considerations concerning relationships between producers and consumers in the production, evaluation, and dissemination of

COMINT. The principals in these discussions and drafting sessions were W. Park Armstrong, State; Loftus E. Becker, CIA; General Magruder, Office of Secretary of Defense; Admiral Wenger, AFSA; and General Canine.[26]

Truman's directive, approved on 24 October 1952, affirmed that communications intelligence was a national responsibility. Truman directed the secretaries of State and Defense as a Special Committee of the NSC for COMINT, to establish, with the assistance of the Director of Central Intelligence, policies governing COMINT activities. He designated the Department of Defense as the executive agent of the government for the production of COMINT information. His memorandum also contained the first reference to the National Security Agency. In addition, Truman's memorandum provided the basis for a reconstitution of USCIB with broadened duties and responsibilities to correspond to those recommended by the Brownell Committee.[27]

On 24 October 1952 the National Security Council issued a parallel document, National Security Council Intelligence Directive Number 9, Revised, entitled "Communications Intelligence." This directive established the new membership of USCIB, defined its duties and responsibilities, and prescribed the procedural methodology governing its participation in COMINT matters. The NSC directive implemented much of Truman's directive.[28]

The composition of the new USCIB differed only slightly from that proposed by the Brownell Committee. Under the NSC directive, the three armed services retained their membership, while the Joint Chiefs were not represented at all. The director, NSA, became a full voting member. The NSC directive also made the DCI the permanent chairman and provided that board decisions should be based upon majority vote. The NSC directive also stated that each member of the board "shall have one vote except the representatives of the Secretary of State and the Central Intelligence

Agency who shall each have two votes."[29] The restructuring of USCIB meant that it became the new mechanism and forum for establishing and adjudicating problems associated with intelligence and processing priorities. With the new voting structure, this assured a more balanced participation by military and civilian representatives of USCIB in the decision-making process.

Simultaneously with the release of NSCID No.9 Revised, President Truman issued a second directive that declared communications security to be a national responsibility to be discharged by a new United States Communications Security Board. He designated the secretaries of State and Defense as a Special Committee of the National Security Council on COMSEC matters and directed them to develop policies and directives relating to the communications security function, including responsibilities, authorities, and procedures.[30]

There remained establishment of the "National Security Agency," as called for by Truman's directive. Secretary Lovett, as the executive agent of the government for communications intelligence, had the basic responsibility for starting the new agency. In the conversion of AFSA to NSA, Lovett had to deal with the issue of communications security. Since Truman's memorandum of 24 October 1952 had excluded communications security from the scope of the COMINT directive, Lovett had to define the extent of the new agency's role in communications security matters. Accordingly, Lovett issued two memoranda associated with the establishment of NSA.[31]

In a remarkably sparse announcement, Secretary Lovett accomplished the actual establishment of the new National Security Agency in his memorandum of 4 November 1952. In his memorandum to the service secretaries, the Joint Chiefs of Staff, and the director, National Security Agency, he described in general terms the basic institutional changes that now governed the cryptologic community. Lovett declared:

- *The designation of the Armed Forces Security Agency was changed to the National Security Agency.*

- *The administrative arrangements for military and civilian personnel, funds, records, and other support categories previously authorized for AFSA were now available and in effect for NSA.*

- *All COMINT collection and production resources of the Department of Defense were placed under the operational and technical control of the director, NSA.*

- *Communications security activities previously assigned to AFSA were now assigned to the director, NSA.*

- *Addressees were directed to appoint a representative to a working group to be chaired by the director, NSA, to develop necessary directives for formal implementation of NSCID No. 9.*[32]

With Lovett's memorandum, Canine acquired a new relationship with the military services and their COMINT activities. Theoretically, he now had control over all COMINT collection and production resources. Canine started a four-year term of office as the director of NSA on 4 November 1952. In accordance with its new charter, he received a third star. Each of the services assigned one or two two-star grade officers to the new agency. This change was consistent with Truman's decision to elevate the status of the unified agency. The reorganization of AFSA removed the COMINT structure

Robert Lovett, secretary of defense

from exclusive military control and theoretically gave all intelligence agencies, military and civilian, an equal voice in the COMINT processing and requirements process.

The new directives, however, did include a "delegation of authority" provision that diluted to some degree the concept of central control. The drafters of the final directives, accepting the Brownell conclusion, supported the position that the services must retain control of the close and direct intelligence support of the forces in the field. Consequently, the final directives made provision for this broad exception by requiring the director to delegate responsibility to the services for direct support as may be required.

The shock waves of reorganization quickly hit the JCS. With the issuance of President Truman's directive and Lovett's follow-up memorandum, the dire predictions about the JCS loss of AFSA had come true. The director, NSA, was no longer under the control of the Joint Chiefs of Staff except, during a transition period, for COMSEC matters. By creating a new agency, Truman had shifted control of COMINT from the JCS to the Department of Defense and the new USCIB. The likelihood of bringing about any reversal of this policy appeared remote.

Three months after the establishment of NSA, Lieutenant General Charles P. Cabell, USAF, director, J-2, presented General Omar N. Bradley, chairman of the JCS, with a lengthy appeal. In Cabell's view, the Brownell Committee had largely overlooked the progress achieved in the postwar evolution of AFSA.[33] Addressing the allegation of AFSA's failure to satisfy the intelligence requirements of State and CIA, Cabell maintained that a major part of the requirements problem had stemmed directly from the failure of State and CIA to seek adjustments in requirements through the existing mechanisms. They had sought a "revolution" instead. Making one final gesture to retain JCS control over the U.S. COMINT effort, Cabell proposed that the secretary of defense delegate responsibility for direction and oversight of the new NSA to the director of the Joint Staff of the JCS. Bradley forwarded the proposal to Lovett, who chose not to override the spirit of the earlier presidential guidance. Instead, Lovett opted to place within his own office the responsibility for exercising a supervisory role over the new NSA. Lovett delegated this responsibility to General Graves B. Erskine, USMC, (Ret.) as the newly established special representative of the secretary's office, who would function without organizational ties to the JCS. This sequence marked the end of JCS efforts to change the direction of the Brownell Report and its implementation.[34]

Established in the fall of 1952, NSA superseded AFSA. The Brownell Committee had succeeded in writing the organizational obituary of AFSA in less than six months. Its report was one of the most significant and far-reaching reviews ever prepared on the COMINT activities of the United States. The recommendations of the committee were accepted and put into effect almost in toto, resulting in a major restructuring of the COMINT community. The report became a kind of Magna Carta for U.S. COMINT activities and the new NSA. Within four months of its completion, a chain reaction of new national-level issuances followed that affected the entire COMINT structure and produced a new COMINT agency.

In summary, there probably never existed a more propitious time for making fundamental changes in the U.S. COMINT structure than in 1951 when the Brownell Committee came into being. After a six-year postwar period of self-study and organizational experimentation, the COMINT community was still groping for answers to a number of major questions. There was also a strong new DCI, who was determined to strengthen the role and mission of his young agency and to establish its permanent niche in the national intelligence structure. The Korean War, in highlighting the intelligence failures of the entire U.S. community, revealed that a great amount of discord and turmoil existed in the

intelligence structure and provided an example for the voices that clamored for fundamental change.

The Brownell Committee, established to conduct an unprecedented wartime probe of the COMINT community, was well staged and well directed by officials from State and CIA. Within six months it produced an impressive report in support of centralization and unification. It radically altered the existing U.S. COMINT structure and permitted U.S. military officials little time to counter its major recommendations.

In the long struggle between the military and civilian agencies over the control of COMINT resources, the turning point came when DCI Smith orchestrated the founding of the Brownell Committee. With Truman's approval, it began its work without the foreknowledge or participation of the military community. Its primary committees and its supporting staff operated without representation from the military community. From the outset, the CIA and State Department dominated the Brownell Committee. There were old animosities to resolve, and CIA and State, while undoubtedly motivated by national security considerations, left nothing to chance in their efforts to realign the COMINT structure and ensure their greater participation in the intelligence process.

Operationally, the Brownell Committee went as far as it could in its proposals for the centralization of COMINT resources. Because of the almost total dependence upon the military installations for the intercept of traffic, the committee concluded that complete centralization of COMINT would not be possible. It recognized that the COMINT services would have to be incorporated into the new centralized structure. Despite this difficulty, the Brownell Report strongly recommended the establishment of a central authority to guide the activities of the military COMINT organizations.

As a result of the Brownell Report, a revised National Security Council Directive of 1952 defined the mission and authority of the new National

Security Agency. NSA remained within the Department of Defense, subject to the direction of the secretary of defense. The director acquired new authorities and responsibilities to assist him in providing unified operational and technical control of COMINT. He acquired operational and technical control over all military COMINT collection and production resources of the United States He was authorized to issue instructions directly to operating units of military agencies engaged in the collection and production of COMINT. The directive did, however, contain one "exception" clause that weakened NSA authority. The directive required that the director make provision for delegation of operational control of COMINT activities to the military services for direct support purposes, as he deemed appropriate. This supported the traditional military position regarding tactical COMINT.

The committee recognized that this exception to the director's control authority would further weaken the concept of centralized control. But in creating a gray area between the services and the central authority, it somewhat optimistically concluded that a solution could be found by the development of greater cooperation between the director and the field commanders.[35]

The implementation of the Brownell Report clearly represented a strong positive move toward unification of the COMINT effort. Because of factors associated with the organizational nature of the military structures, there remained the same number of agencies engaged in cryptologic activities. But NSA represented a vastly stronger structure than AFSA. With the acknowledgment by Brownell that direct support to field forces should be controlled by the Service Cryptologic Agencies rather than NSA, the services retained a significant degree of independence. This still presents problems today.

Chapter VII

Summary: The Struggle for Control Continues

World War II marked a key juncture in the growth and expansion of the U.S. Army and Navy COMINT operations. The war had a monumental and immediate impact on U.S. COMINT collection and reporting operations. Even more significantly, World War II marked the start of ten years of change and evolution for the military cryptologic services. After World War II, there was no returning to the era of total independence for the military COMINT organizations. Massive changes culminated with the establishment of the National Security Agency in 1952 and the strengthening of the United States Communication Intelligence Board. While the organizational origins of the new agency represent a fairly simple audit trail, the political struggles and cross-pressures that led to the establishment of NSA are far more complex. The expanding intelligence requirements of the federal government, the passage of the National Security Act of 1947, budgetary considerations, and bureaucratic infighting between the military and civilian agencies all were prominent factors in the effort to centralize the communications intelligence functions of the federal government into one agency.

By the summer of 1942, as a result of action by President Roosevelt, the Army and Navy cryptologic structures became the principal U.S. organizations devoting efforts to foreign communications intelligence activities. Their organizations had evolved along different lines, within different departments, and no one organization directly supervised their efforts. As a result of this dichotomy of origins and structure, a well-established pattern of independence – if not isolation – characterized Army-Navy relationships on COMINT matters. In June 1942 the services did reach an agreement on a division of

cryptanalytic tasks, but there occurred no immediate change in their working relationships.

Until 1942 the Army and Navy resisted the introduction of any major changes to their relationships and sought to maintain their traditionally separate cryptanalytic roles. Each worked independently and exclusively on its assigned cryptanalytic tasks. The services not only continued to demonstrate little enthusiasm toward closer cooperation in COMINT matters, but maintained their traditional hostility toward proposals for merger, or even of opening up a new dialogue on operational problems. Consequently, cooperation on COMINT matters was minimal during the first two years of the war.

Nevertheless, out of the disaster at Pearl Harbor and the pressures of all-out war came persistent demands for the establishment of a truly centralized, permanent intelligence agency. As early as 1943 proposals for the establishment of a single United States Intelligence Agency routinely surfaced in the various intelligence forums of the JCS. At the same time, some military COMINT authorities foresaw their vulnerability to congressional criticism and future reductions in resources since they conducted their COMINT operations on a fractionated and sometimes duplicatory basis. Recognizing these threats to a continuation of their separate existence, the Army-Navy COMINT organizations took steps to establish closer technical cooperation.

In 1944 some positive signs of the services moving toward an expansion of interservice cooperation occurred. That year saw the conclusion of a number of technical agreements between the services and the first exchange of liaison officers in Washington on a number of certain problems.

It also saw the establishment of ANCICC. While the dialogue was carefully prescribed and did not change the overall independent operations, it represented movement toward some form of consolidation.

Operating as separate COMINT organizations, both the Army and Navy experienced major successes during the war. Included were cryptanalytic breakthroughs against the communications of German submarines and armed forces. These accomplishments heightened the sense of value and appreciation of intelligence among the military commanders and the leaders or the government. Communications intelligence generally came to be identified as the most important form of intelligence. Ironically, the magnitude or these intelligence successes later became the measuring rod for criticism of the postwar achievements of the military COMINT organizations.

As the war came to a conclusion, some Army and Navy officials realized that the loss of their primary targets meant dire consequences for their organizations and budgets. Because of worsening Soviet-U.S. relations, however, the services began to explore the possibility of directing a major effort against Soviet targets. The services also anticipated major organizational changes in intelligence activities as the war wound down. The first of these changes occurred within a few months after V-J Day when President Truman ordered the establishment of new intelligence organizations, and authorized continuing relations with the British cryptologic organization. In January 1946 Truman created a National Intelligence Authority, a Central Intelligence Group, and a Director of Central Intelligence.

Eighteen months later, Congress passed the National Security Act of 1947, which reinforced and amplified the earlier Truman action concerning centralization of the U.S. intelligence effort. The act gave birth to a National Security Council, a Central Intelligence Agency, and a National Military Establishment, with three coequal departments of

the Army, Navy, and Air Force. By 1948 a third military COMINT organizatioj emerged, the Air Force Security Service. It began competing for scarce COMINT targets and resources.

Within a few months after the Truman Directive of 1946, the service COMINT organizations initiated their own reorganization effort. This effort marked the beginning of six years of experimentation. Basically, the military authorities sought to centralize military control over COMINT activities and to develop an organization that would be responsive to military needs, especially with regard to the Soviet Union.

The first major change occurred in May 1946 when the services formed a joint working agreement, which became known as the Joint Operating Plan. The plan brought about a voluntary collocation of Army and Navy processing activities in the United States. Under the JOP, however, the services retained their separate identities and organizations. The plan also called for a radically new position, the Coordinator or Joint Operations. The position was literally that – a coordinator, not a director of operations.

Under the new JOP a new layer of committees subordinate to the CJO was also created. The CJO became a super-chairman for all the committees established under the JOP. Although he had a coordination role, he was powerless to direct the services, even on matters of joint tasks. This management weakness was compounded further when Army-Navy officials failed to reach agreement on what constituted joint tasks or the amount of their manpower contributions to joint tasks. Moreover, by this time, the civilian agencies had come to recognize that they had little or no voice in setting intelligence priorities for COMINT. Military interests simply dominated the process.

In late 1947 a major struggle developed between the military and civilian members of USCIB. Admiral Roscoe Hillenkoetter, the third DCI, became the primary catalyst for the issuance of a new charter for USCIB. Hillenkoetter's general intent was to rewrite the charter to reflect the expanded membership of USCIB and to correlate the authorities of the Communications Intelligence Board with the National Security Act of 1947. Hillenkoetter wanted to give the civilian agencies a greater voice on policy matters relating to COMINT. He openly sought to bring U.S. COMINT under the direct control of the DCI.

After several months of negotiations, the members of USCIB (Army, Navy, Air Force, State, and CIA) could not agree on which organization should have the ultimate authority over the COMINT community. (The FBI retired from the board in 1947.) The board was deadlocked. The armed services took the position that USCIB should report to the Joint Chiefs of Staff. State and, however, believed that the board should report stalemated questions to the National Security Council instead.

On 1 July 1948 the National Security Council broke the deadlock by issuing National Security Council Intelligence Directive No.9, "Communications Intelligence." The new directive, with the strong personal support of the secretary of defense, James Forrestal, represented a major victory for the civilian members of USCIB. Under the provisions of the new NSCID No.9, USCIB reported to the NSC as its parent body rather than to the individual military department heads. Also, for the first time, USCIB had an official charter issued at the national level. The rule of unanimity continued to govern USCIB's decision-making process, however, and hindered the effective functioning of the board.

Although Hillenkoetter achieved a major victory with the issuance of NSCID No. 9, he failed in his attempt to place the COMINT functions directly under the DCI. In the view of most military authorities, however, the outcome was still a catastrophe. The JCS clearly lost out in its counterproposal to be designated the "parent body" of USCIB for unresolved issues. Nevertheless, while NSCID No.9

effectively dealt the JCS a blow in its efforts to control U.S. COMINT activities, it did not result in any immediate change in the day-to-day activities of USCIB and its subordinate committees. Since the military organizations had a majority on the board, they continued to dominate the discussions. The situation was clearly changing, however. As the major beneficiaries of the new directive, the State Department and CIA began to exert a much greater influence in all COMINT deliberations and decisions.

Within ten months of the issuance of the NSC directive, another major change took place in the intelligence structure. On 20 May 1949 Defense Secretary Louis Johnson directed a physical merger of the central processing activities of the three cryptologic services by establishing the Armed Forces Security Agency. He placed the cryptologic functions under the exclusive control of the Joint Chiefs of Staff. AFSA came about as a result of two interrelated political factors. On the one hand was the announced objective of Secretary Johnson to achieve "efficiency and economy" in the management of the cryptologic effort. On the other was the obvious strategy of the JCS to strengthen and to reestablish its hold on COMINT resources. The military was once again in a dominant position on COMINT matters.

From the perspective of the civilian agencies, the creation of AFSA meant the renewal of the military and civilian struggle over the control of COMINT resources. CIA and State Department representatives argued strongly against AFSA, which in their view existed in direct conflict with the new USCIB charter. They maintained further that AFSA was established without their participation and over their protests. Secretary Johnson, however, not only refused to discuss it directly with them, but refused as well to make any changes in the basic AFSA charter. Johnson did make one concession. He canceled the proposed Armed Forces Communication Intelligence Board, which would have become a policy board running parallel to USCIB.

Although the establishment of the Armed Forces Security Agency seemed to represent a consolidation of the U.S. COMINT effort and a more efficient approach to U.S. COMINT activities, AFSA was fundamentally unsound from both a conceptual and managerial viewpoint Pentagon authorities, however, viewed AFSA as a reasonable and evolutionary step toward "service unification." Unification proved to be an ephemeral and elusive concept, however.

Although some military officials acquiesced in the concept of consolidation, it soon became clear that the bureaucracy in each service never seriously envisioned a true merger and the resulting diminution of its own responsibilities and authorities. AFSA was the creation of Louis Johnson, secretary of defense. He sought to achieve a degree of unification of the services as well as "efficiency and economy" in the management of the cryptologic structure. While a form of merger took place, no fundamental changes were made in the way each service conducted its own operations.

In the actual implementation of the AFSA charter, the services took full advantage of loopholes in the charter to preserve their independent status. For example, the Air Force used the "exclusion clause" in AFSA's charter (which withheld from it any authority for the tasking of mobile collection sites) to exclude AFSA from any role in controlling Air Force collection sites. In fact, by 1952, AFSA had no authority over any Air Force collection sites. All had been conveniently identified by the Air Force as mobile facilities. In addition, two months after forming AFSA, the Joint Chiefs made substantive changes in the AFSA charter, and drastically diluted its basic authorities.

These problems, combined with a waning military support for the general AFSA concept, foretold its ultimate demise. Of the three COMINT services, it was ironic that the Navy, which from the outset had strongly opposed even the AFSA concept of cooperation, ultimately provided the greatest support for AFSA. Although the Army, in the person of

Colonel Carter W. Clarke, became identified as the originator of the AFSA concept, Army support for its offspring quickly diminished and could be characterized at best as lukewarm. The Air Force, with its newly established AFSS, aggressively opposed AFSA, seeking primarily to build its own structure and achieve total independence in COMINT matters.

From AFSA's first days, there was no way in which its first director, Admiral Earl Stone, could make it operate as a centralized unified structure. However, the full extent and impact of the weaknesses of the AFSA charter would not become widely known – or recognized – until the onset of the Korean War. By 1951 General Ralph Canine, the second director, AFSA, was encountering the same open opposition from the services to his efforts at centralization and consolidation as Stone had experienced.

The Korean War revealed the inherent weaknesses not only of the AFSA structure, but of the USCIB as well. During the Korean War, U.S. COMINT produced a mixed record. Its major successes took place in the area of tactical support, achieved primarily by the Army and Air Force. AFSA came under heavy criticism because of its problems in attempting to control and direct the military COMINT services. The civilian members of USCIB now pointed to the lack of direction and unity in the COMINT effort. AFSA was not providing results.

The Army proposal to establish a new military agency to be known as the Consolidated Special Information Dissemination Office (CONSIDO) shocked the civilian members of USCIB. The proposal drastically limited civilian input on COMINT matters. CONSIDO soon became the symbol of a new battle between the military and civilian members of USCIB for control of COMINT.

CIA and State officials completely understood the military rationale for the establishment of AFSA – although they thoroughly disagreed with it. The

CONSIDO proposal, however, represented an even more encompassing threat to civilian input to the COMINT process. Not only was the control of COMINT and its dissemination at stake, but the control of all-source intelligence estimates and evaluation actions appeared to be at risk as well. The concept of a military CONSIDO controlling dissemination, estimates, and evaluative actions seemed to crystallize the major fears of the civilian agencies about their diminishing policy role in all intelligence matters. Because of this major opposition, the proposal died in USCIB in December 1950.

The long debate over CONSIDO left a lasting impression on CIA, State, and the FBI (reinstalled as a member of USCIB in 1950). They now believed that the military authorities would not relent in their pursuit of the CONSIDO-type concept, and would probably submit an amended version of CONSIDO at a later date.

By 1951 it was clear to the civilian agencies that the military organizations were incapable of jointly developing a structure that would meet, without bias, the needs of the growing United States intelligence community. After six years of experimentation and reorganization and two attempts to consolidate and centralize the communications intelligence activities of the United States, instability, disunity, and decentralization still existed. and State were not totally altruistic in their opposition to military plans. They often appeared more concerned about the long-range overtones of military control of the intelligence role than about the actual level of COMINT support received during the Korean War, for example. They realistically concluded, however, that any fundamental reworking of the communications intelligence structure would come about only as a result of outside intervention. Any further joint military-civilian dialogue seemed useless. Working together, CIA and State officials proceeded to develop their own strategy for a "new look" at the organization of the COMINT structure. This time the civilians would have a major input.

The military authorities previously had set up AFSA without prior coordination with USCIB or the civilian members of the COMINT community. Now a complete reversal took place. The military authorities were completely left out of the deliberative and decision-making process leading to the termination of AFSA and the creation of a new centralized COMINT agency.

General Walter Bedell Smith, as the fourth DCI, became the catalyst for bringing about a new national-level review of the COMINT structure. In a memorandum to the National Security Council, dated 10 December 1951, Smith recommended an overall review of United States COMINT activities, based upon an earlier study by Kingman Douglass. The NSC, in turn, forwarded the proposal to President Truman. Three days later, on 13 December 1951, Truman directed Secretary of State Dean G. Acheson and Secretary of Defense Robert A. Lovett, assisted by Director of Central Intelligence Smith, to review in depth the communications intelligence activities of the United States. The resulting review process was carefully orchestrated.

On 28 December 1951, in response to Truman's request, Acheson and Lovett established the Brownell Committee to study the existing structure and make recommendations. George A. Brownell, an eminent attorney in New York City, headed the committee. Brownell served as chairman, assisted by Charles E. (Chip) Bohlen, counselor, State Department; William H. Jackson, special assistant to the DCI; and Brigadier General John Magruder, USA (Ret), special assistant to the secretary of defense. The CIA and the Department of State provided the four staff members for the committee, all of whom had served previously in the special intelligence branches of the Army or Navy. During the period of the survey, the Brownell Committee and its support staff resided at CIA and received administrative support from the CIA. The military organizations had no representation on the Brownell Committee or on its support staff.

Within six months, the Brownell Committee completed its report. It stressed the need for the unification of U.S. COMINT responsibilities and recommended a major overhaul of the existing COMINT organization as well as the USCIB structure. The final Brownell Report completely demolished the concept of "unification" as it existed under AFSA. During the next four months, extended negotiations took place among the representatives of CIA, Departments of State and Defense, and the director, Armed Forces Security Agency, over the exact wording of the implementing directives to be issued by the president. The Joint Chiefs of Staff were also excluded from these discussions. Ten months after the establishment of the Brownell Committee, Truman, accepting the report, issued two directives that led to the establishment of the National Security Agency with its dual responsibility for the communications intelligence and communications security activities of the government. There would be a centralized authority for U.S. COMINT activities, and the civilian authorities, by virtue or a major restructuring of USCIB, would play a major role in directing the scope of NSA's operations.

In conclusion, the directive establishing NSA clearly identified the national rather than the solely military character of U.S. COMINT activities. Within well-defined limits it strengthened the role and authorities of the director, NSA, over COMINT. It greatly expanded administrative and operational controls over all U.S. cryptologic activities. For the first time, the director acquired the authority to issue instructions directly to military units without going through military command channels. However. there remained some significant built-in limitations in the NSA charter. Although a nominal unification took place, efforts to unify and centralize COMINT authorities and responsibilities in one organization achieved only a partial and limited success.

From the outset, the designers of the NSA charter clearly recognized that complete unification would be impossible because of the dependence upon the military structures to man field stations. Consequently, although the service units were incorporated organizationally into the central organization. they retained their own authorities and responsibilities within their military departments. To ameliorate this, and to assist NSA in its mission of providing effective unified organization and control of COMINT, the enabling directives provided that NSA would have operational and technical control over all the COMINT collection and production resources of the United States. Even this did not solve the problem.

There also existed a "delegation of authority" clause in the new charter that further diluted the concept of centralized control. The Brownell Committee, as well as the drafters of the implementing presidential directive, supported the services' traditional position that they must control the close and direct intelligence support of the forces in the field. Consequently, the final directive made provision for this broad exception by requiring the director to delegate responsibility to the services for direct support as may be necessary. The committee fully recognized that it was creating a problem area between the central authority and the services, but concluded that a solution could be found "with sincere and intelligent cooperation between the commanders involved."

Finally, despite the reorganization the same number of agencies remained engaged in cryptologic activities as before – namely, NSA, CIA, Army, Navy, and Air Force. NSA had in many respects simply replaced the defunct AFSA. The services retained a significant degree of independence. They retained their own separate organizations and identities, as well as administrative and logistic control of their field operations. The struggle over who would control U.S. COMINT resources would continue.

Abbreviations

AFCIAC–Armed Forces Communications Intelligence Advisory Council (later redesignated as Armed Forces Security Agency Council)

AFCIB–Armed Forces Communications Intelligence Board

AFOIN–Air Force Office of Intelligence

AFSA–Armed Forces Security Agency

AFSAC–Armed Forces Security Agency Council

AFSG- Air Force Security Group

AFSS - Air Force Security Service

ANCIB–Army-Navy Communications Intelligence Board (joint policy board that later became State-Army-Navy Communications Intelligence Board)

ANCICC–Army-Navy Communications Intelligence Coordinating Committee (working level committee of Army-Navy Communications Intelligence Board)

ASA–Army Security Agency

ASAPAC–Army Security Agency Pacific

CAHA–Cryptologic Archival Holding Area

CIA–Central Intelligence Agency

CIG–Central Intelligence Group

CJCS–Chairman, Joint Chiefs of Staff

CJO–Coordinator of Joint Operations

CNO–Chief of Naval Operations

COI–Coordinator of Information

COMINCH–Commander in Chief, U.S. Fleet

COMINT-Communications Intelligence

COMMSUPDETS–Communications Supplementary Detachments

COMSEC–Communications Security

CONSIDO–Consolidated Information Dissemination Office

CRB–Communications Reconnaissance Battalion

CRC–Communications Reconnaissance Company

CRG–Communications Reconnaissance Group

CSAW–Communications Supplementary Annex, Washington (Navy facilities at Ward Circle, Washington, D.C.)

DCI–Director of Central Intelligence

DIA–Defense Intelligence Agency

DJO–Director of Joint Operations

DJS–Director, Joint Staff

DoD–Department of Defense

DOE–Department of Energy

ELINT–Electronic Intelligence

E.O.–Executive Order

FBI–Federal Bureau of Investigation

FCC–Federal Communications Commission

FECOM–Far East Command

G.C. & C.S.–British Government Code and Cipher School

GCHQ–Government Communications Headquaters (U.K. SIGINT Organization)

G-2–Intelligence Division, War Department General Staff

IAB–Intelligence Advisory Board

IAC–Intelligence Advisory Council

JIC–Joint Intelligence Committee (within the Joint Chiefs of Staff structure)

JICG–Joint Intercept Control Group

JCS–Joint Chiefs of Staff

JLG–Joint Liaison Group

JN-25–U.S. Navy designator for Japanese 5-digit code used by Japanese fleet

JOP–Joint Operating Plan of Army and Navy, 1946-1949

JPAG–Joint Processing Allocation Group

LSIB–London Signals Intelligence Board

NIA–National Intelligence Authority

NME–National Military Establishment

NSA–National Security Agency

NSC–National Security Council

NSCID–National Security Council Intelligence Directive

NSS–Naval Security Station

ONI–Office of Naval Intelligence

OPNAV–Office of Naval Operations

OSD–Office of the Secretary of Defense

OSS–Office of Strategic Services

RAF–Royal Air Force

SCA–Service Cryptologic Agency

SECNAV–Secretary of Navy

SIGINT–Signals Intelligence

SIS–Signal Intelligence Service (Army)

SMP–Special Committee on Merger Planning

SRB–Special Research Branch (Army Intelligence)

SSA–Signal Security Agency (Army)

STANCIB–State-Army-Navy Communications Intelligence Board (joint policy board that later became the United States Communications Intelligence Board)

STANCICC–State-Army-Navy Communications Intelligence Coordinating Committee (working level committee of State-Army-Navy Communications Intelligence Board)

SUKLO–Senior U.K. Liaison Officer

SUSLO–Senior U.S. Liaison Officer

SWI–Special Weather Intelligence

SWNCC–State-War-Navy Coordinating Committee USA-U.S. Army

USAF–U.S. Air Force

USCIB–United States Communications Intelligence Board

USCICC–United States Communications Intelligence Coordinating Committee (working level committee of United States Communications Intelligence Board)

USEUCOM–U.S. European Command

USN–U.S. Navy

WC–War Council

WDC/CCO–Washington, D.C., Control-Collection Office (Air Force Security Service)

Notes

Chapter 1

1. Admiral Joseph R. Redman, OP-20, memorandum for the Secretary of the Navy. "The Navy's Interest in the Processing of Intercepted Foreign, non-military Communications," 24 September 1945 Cryptologic Archival Holding Area, Accession No. 4614, CDIB 38, NSA (TS). (Hereafter cited as Redman memorandum, 24 September 1945.)

2. *History of the Signal Security Agency in World War II*, "Organization, 1939-1945," (ASA, 1945), Vol. I, Part I, 90-103, History Collection, NSA, Series IV (TS). (Hereafter cited as *History of the Signal Security Agency*, Vol. I, Part 1.)

3. John W. McClaran, OP-20-G, memorandum for the Director, 12 April 1933, (Documents on Pre-1952 Cryptologic Organization and Policy, Vol. 1, (Pre-1942)), History Collection, NSA, Series VII (s).

4. The Navy had the primary responsibility for naval radio activities relating to all foreign powers, and for the diplomatic radio intelligence activities of the four major naval powers (England, France, Italy, and Japan). The Army had the primary responsibility for military radio traffic relating to all foreign powers, and for the diplomatic traffic of all foreign powers other than England, France, Italy, and Japan.

5. Laurance F. Safford, OP-20-G, memorandum for OP-20, "Coordination of Intercept and Decrypting Activities of the Army and Navy," 25 July 1940, Cryptologic Archival Holding Area, Accession No. 4614, CBID 38, N5A (5). (Hereafter cited as Safford memorandum, 25 July 1940)

6. Ibid.

7. Commander Laurance F. Safford and Colonel Spencer B. Akin, Directive to Joint Army-Navy Committee, 31 July 1940, Cryptologic Archival Holding Area, Accession No. 4614, CBIB 38, NSA (S).

8. Commander Laurance F. Safford, OP-20-G, memorandum for Admiral Leigh Noyes, 27 July 1940, Cryptologic Archival Holding Area, Accession No. 4614, CBIB 38, NSA (S).

9. Safford memorandum, 25 July 1940.

10. Ibid.

11. John R. Redman, OP-20-G, memorandum to the Vice Chief of Naval Operations, "Cryptanalytical and Decryption Operations on Diplomatic Traffic," 25 June 1942, Cryptologic Archival Holding Area, Accession No. 4614, CBIB 38, NSA(S); and Laurance F. Safford, memorandum for OP-20-G, "Responsibility for Decoding and Translating Japanese Intercepts," 14 February 1946. History Collection, NSA, Special Series VII, Box 4. (Hereafter cited as Redman memorandum, 25 June 1942 and Safford memorandum, 14 February 1946.)

12. Ibid.

13. Colonel Spencer B. Akin and Commander Laurance F. Safford, memorandum for Chief Signal Officer and the Director of Naval Communication, Traffic Division, 3 October 1940, History Collection, NSA.

14. Ibid.

15. Safford memorandum, 14 February 1946.

16. Redman memorandum, 24 September 1945.

17. Joseph N. Wenger, Vice Director, NSA, memorandum for Grant Manson, Staff Member, Brownell Committee, "Draft of Part I of the Brownell Committee Report," 27 March 1952, History Collection, NSA, Series VI.C (TSC); *History of the Signal Security Agency*, Part I, 90.

18. Frederick D. Parker, *A New View to Pearl Harbor: United States Navy Communication Intelligence, 1924-1941*, United States Cryptologic History Series, (NSA, 1988), Series IV, World War 11. Vol. II, 53-55,1988 (s).

19. Captain John R. Redman, memorandum for Captain Joseph N. Wenger, (OP-20-G), "Army-Navy Agreement Concerning Allocation of Diplomatic Traffic," 9 February 1945, Cryptologic Archival Holding Area, Accession No. 4614, CBIB 38, NSA(TS).

20. Ibid.

21. Redman memorandum, 25 June 1942.

22. Safford memorandum, 25 July 1940.

23. Ibid.

24. Redman memorandum, 25 June 1942; and Joseph N. Wenger, memorandum for OP-20-G, "Future Cooperation between Army and Navy," 1 June 1943, Cryptologic Archival Holding Area, Accession No. 4614, CBIB 38, NSA(S).

25. Redman memorandum, 25 June 1942.

26. Joseph N. Wenger, memorandum for OP-20-G, "Future Cooperation between Army and Navy," 1 June 1943, Cryptologic Archival Holding Area, Accession No.

4614, CBIB 38, NSA (S); and History of Signal Security Agency, World War 11, 93.

27. Report of the Army-Navy-FBI Allocation Committee, "Report of Conference Appointed to Study Allocation of Cryptanalysis," 30 June 1942, Cryptologic: Archival Holding Area, Accession No. 4614, CBIB 38, NSA (s). (Hereafter cited as Report of Allocation Committee, 30 June 1942.)

28. Ernest J. King, Commander in Chief, U.S. Fleet and George C. Marshall, Chief of Staff, memorandum for the President, 6 July 1942, Cryptologic Archival Holding Area, Accession No. 4609, CBlB 38, NSA(S).

29. Franklin D. Roosevelt, memorandum for the Director of the Budget. 8 July 1942, Cryptologic Archival Holding Area, Accession No. 4609, CBIB 38, NSA(S)

30. Redman memorandum, 25 September 1945.

31. Report of Allocation Committee, 30 June 1942. A summary of the committee's allocation for cryptanalyis follows:

Diplomatic:	Army
Enemy Naval Operations:	Navy
Enemy Military Operations:	Army
Western Hemisphere Clandestine:	FBI-Navy
International,Clandestine (other than Western Hemisphere):	Navy
Trade Codes:	(To be assigned by Committee)
Army Weather:	Army
Navy Weather:	Navy
Domestic Criminal:	FBI
Voice Broadcast:	FBI
Cover Text Communications:	FBI
Miscellaneous:	(To be assigned by Committee)

32. Report of Allocation Committee, 30 June 1942.

33. History of Signal Security Agency, Vol. I,Part 1, 121.

34. Ibid.

35. The NAVSECGRU Bulletin, Vol III, Number 2, March-April 1969 History Collection, NSA. Series XII.H (U).

36. OP-20-G "Outline of the Collaboration in Cryptanalysis Between the Army and the Navy," 18 August 1944, 7, Cryptologic Archival Holding Area, Accession No. 4609, CBIB 38. (Hereafter cited as Outline of Collaboration by Army and Navy, August 1944.)

37. Outline of Collaboration by Army and Navy, August 1944, 8 (TS).

38. George C. Marshall and Ernest J. King, "Joint Army-Navy Agreement for the Exchange of Communications Intelligence" 4 February 1944, Cryptologic Archival Holding Area, Accession No. 2465, CBIB 14, NSA (S). (Hereafter cited as Joint Army-Navy Agreement, 4 February 1944.)

39. SRH-200, OP-20-G File on Army-Navy Collaboration, 1831-1945, Part 2, 7, Special Research History (SRH) 200; Record Group 457, National Archives. (Hereafter cited as RG-457, NSA.)

40. Briefing by Brigadier General Woodbury M. Burgess, USA, Assistant Director for Production, February 1957, File: Robertson Committee Background, History Collection NSA, Series VI.C (8).

41. History of the Signal Security Agency, "The Japanese Army Problem - Cryptanalysis," 13 May 1947, Vol. 3, 24.

42. Henry F. Schorreck. The Role of COMINT in the Battle of Midway, SRH 230, 00-457.

43. Brownell Committee, Report to the Secretary of State and the SecDef, 13 June 1952, 22-29. History Collection. NSA, Series V.F (TSC).

44. Admiral Joseph R. Redman, OP-20, memorandum for Admiral Edwards, "Formalization of Army-Navy Communication Intelligence Coordinating Committee (ANCICC)–Need for," 18 January 1940, Cryptologic Archival Holding Area, Accession No. 2463, CBIB 14, NSA (TS). (Hereafter cited as Formalization of ANOCC, 18 January 1945.)

45. Joseph N. Wenger, OP-20-G, memorandum for OP-20 via F20, Liaison," 16 July 1945, Cryptologic Archival Holding Area, Accession No. 7779, CBQM 36, NSA (U).

46. Dennis DeBrandt, "Structuring Intelligence For War," CIA Studies in Intelligence, Vol. 32, No.1, Spring 1988, 43-56 (U).

47. James S. Lay, CIA, draft manuscript, undated, entitled "History of USCIB," Vol I, 3, History Collection, NSA, Series XII.H. Box 47 (8). (Hereafter cited as Lay Report.)

48. Ibid.

49. During this period, intelligence requirements were generally handled on a service-to-service basis. For

example, the military consumers would originate their intelligence requirements and forward them to the respective COMINT organizations for satisfaction. Following the establishment of USCIB, and the passage of the National Security Act of 1947, minor procedural changes in the process started to take place in the post-war years. The first major effort to change the requirements process occurred in the mid-1960s with the establishment of the Intelligence Guidance for COMINT Programming (IGCP). Further changes took place in the mid-1970s when the intelligence community established a National SIGINT Requirements System (NSRS) that superseded the IGCP.

50. Lay Report, Vol. 1, 5.

51. In 1941 New York lawyer William J. Donovan was appointed Coordinator of Information (COI) by President Roosevelt. In 1942 this organization was redesignated the Office of Strategic Services and transferred to the jurisdiction of the Joint Chiefs of Staff. Its mission was to collect and analyze strategic information as required by the Joint Chiefs of Staff and to conduct special operations not assigned to other agencies. *United States Government Organization Manual,* Summer 1944, 162. See also John Ranelagh, *Thc Agency: The Rise and Decline of thc CIA* (New York: Simon and Schuster, 1986).

52. Formalization of ANCICC, 18 January 1945.

53. Ibid.

54. Joseph N. Wenger, memorandum for the Chairman, USCIB, Report on the Status and Authority of USCIB, USCICC, its members and the Coordinator of Joint Operations," 7 February 1947, Appendix A, 1, Cryptologic Archival Holding Area, Accession No. 5061, CBIB 23, NSA (TS). (Hereafter cited as Wenger Report to USCIB, 7 February 1947.)

55. Minutes of third meeting of Army-Navy Communication Intelligence Coordinating Committee, 19 June 1944, Cryptologic Archival Holding Area, Accession No. 2154, CBPC 26, NSA (TSC).

56; General George C. Marshall, memorandum for Admiral Ernest King, "Army-Navy Communication Intelligence Board–Establishment of," 9 March 1945, Cryptologic Archival Holding Area, Accession No. 2463, CBIB 14, NSA (TS).

57. Wenger Report to USCIB, 7 February 1947, Annex C-1.

58. Wenger Report to USCIB, 7 February 1947, Annex C-2.

59. Formalization of ANCICC, 18 January 1945.

60. Ibid.

61. ANCIB was primarily structured to serve as a coordinating mechanism for handling the joint plans and operations of the Army and Navy COMINT organizations, and such other issues as could be resolved amicably. From the outset, however, ANCIB was intended to function only as a brokering or coordinating organization, and not as an authoritarian voice. While the COMINT services accepted ANCIB as a coordinating and negotiating mechanism, they never viewed ANCIB's charter as giving it sufficient authority to cut across the command channels of the Army and Navy whenever disagreements arose. The heads of the Army Security Agency and the Navy Supplementary Activity continued to be directly responsible to their respective military commanders.

62. Dean Acheson, Henry L. Stimson, and James V. Forrestal, joint memorandum for President Truman, "Collaboration with the British in the Communication Intelligence Field, Continuation and Extension of," undated, History Collection, NSA., Series V.A (TSC). (Hereafter cited as Acheson, Stimson, and Forrestal memorandum.)

63. Richard Hough, *The Greatest Crusade - Roosevelt, Churchill, and the Naval Wars* (New York: William Morrow and Company, Inc. 1986), 25-27.

64. *History of the Signal Security Agency*, Vol. 1, Part I, 144.

65. Lieutenant Commander Alwin D. Kramer, memorandum for OP-20-G, "Cryptanalysis: FBI Activities and Liaison with the British," 8 June 1942, History Collection, NSA Series XII.H. Box No. 47 (S).

66. Franklin D. Roosevelt., memorandum for General Marshall, dated 9 July 1942. Text of memorandum is quoted in *History of the Signal Security Agency in World War II*, Vol. 1. Part I, 118.

67. Ibid.

68. Major General George V. Strong, Assistant Chief of Staff, G-2, memorandum for Chief of Staff, 9 July 1942. Text of memorandum is quoted in *History of the Signal Security Agency*, Vol. 1, Part I, 110.

69. Lieutenant General Joseph T. McNarney, USA, Deputy Chief of Staff', memorandum for the Chief of Staff, " Agreement between British Government Code

and Cipher School. and U.S. War Department in regard to certain Special Intelligence," 10 June 1943, Cryptologic Archival Holding Area, Accession No. 2401, CBTJ 66, NSA(S).

70. Outline of Collaboration by Army and Navy, August 1944, 44-48.

71. Formalization of ANCICC, 18 January 1940.

72. Ibid.

73. Minutes of 27th meeting of the State-Army-Navy Communication Intelligence Coordinating Committee, 13 February 1946, History Collection, NSA. Series XII.H, Box 37 (TS).

74. Rear Admiral Hewlett Thebaud, Chairman,ANCIB, memorandum for General Clayton Bissell, Admiral John R. Redman, and General W. Preston Corderman, "British Reply to U.S. Proposal" 15 August 1945, Cryptologic Archival Holding Area, Accession No. 7779, CBQM 36, NSA (TSC). (Hereafter cited as British Reply.)

75. John F. Callahan and Wason G. Campbell, Secretariat of STANCIB/STANCICC, memorandum for members of STANCICC, "Establishment of a U.S. Combined Intelligence Liaison Center in Great Britain," 3 May 1946, Cryptologic Archival Holding, Area, Accession No. 2156, CBPC 27, NSA(S).

76. Ibid.

77. Acheson, Stimson, and Forrestal memorandum.

78. Harry S. Truman, memorandum for Secretaries of State, War, and Navy, 12 September 1946, History Collection, NSA. Series V.A (TS).

79. Minutes of the 22nd meeting of Army-Navy Communication Intelligence Coordinating Committee, 16 October 1945, Cryptologic Archival Holding Area, Accession No. 2154, CBPG 26, NSA(TSC).

80. Rear Admiral Joseph R. Redman, Senior Member, ANCIB, memorandum for General Eisenhower and Admiral King, 13 December 1945. "State Department Representation on ANCIB-ANCICC," Cryptologic Archival Holding Area. Accession No. 5061, CBIB 23, Annex 0-2, 1, NSA (TSC).

81. Rear Admiral Joseph R. Redman, Senior Member. ANCIB, memorandum for Alfred McCormack, "State Department Representation on ANCIB-ANCICC," 11 December 1945, Cryptologic Archival Holding Area, Accession No. 5061, CBIB 23, Annex 0.2, 2, NSA (TS). Alfred McCormack entered the War Department as special assistant to Henry Stimson, secretary of war in 1942.

He was appointed deputy chief special branch. Military Intelligence Division, War Department General Staff on 17 June 1942. In 1944 he became director of Intelligence, Military Service, in charge of providing intelligence for the Operations and Planning Division and for the command in all theaters of war. On 1 October 1945 McCormack became special assistant to James F. Byrnes, secretary of state, with the task of organizing an intelligence unit in that department. He resigned from State in April 1946.

Chapter II

1. The Navy provided posthumous recognition of Admiral Wenger on 7 July 1976, with the dedication of the Rear Admiral Joseph N. Wenger Naval Cryptologic Museum. In the dedication ceremony, special note was made of his "helping to establish the Armed Forces Security Agency and later serving as Vice Director of its successor, the National Security Agency."

2. Minutes of the 14th meeting of Army-Navy Communications Intelligence Coordinating Committee, 9 May 1945, Section III (Army Proposal Regarding Army-Navy Collaboration), 7-8, Accession No. 2154, CBPC, Cryptologic Archival Holding Area, NSA (TS). Captain J.N. Wenger, OP-20-G, memorandum for Captain W.R. Smedberg, III (F-20). "Policy regarding collaboration between the Army and the Navy in the Communication Intelligence Field," 30 May 1945, Accession No. 5061 N. CBIB 23, Cryptologic Archival Holding Area, NSA (S).

3. Minutes of 8th meeting of Army-Navy Communication Intelligence Coordinating Committee, 8 November 1944, Section (Joint American-British Postwar Agreement), 6-8, Accession No. 2154, CBPC 26, Cryptologic Archival Holding Area, NSA (TS).

4. OP-20-G staff study, "Outline of the Collaboration in Cryptanalysis between the Army and Navy," 18 August 1944, Part VII (Discussions prior to 1944), 55-69, Accession No, 2468, CBIB 15, Cryptologic Archival Holding Area, NSA (TS).

5. Ibid., 58.

6. George C. Marshall, memorandum to Admiral King, "Signal Intelligence," 18 August 1945, Accession No. 5061 N, CBSB 23, Cryptologic Archival Holding Area, NSA (TSC).

7. Fleet Admiral Ernest J. King, memorandum for General George C. Marshall, "Signal Intelligence," 20 August 1945, Enclosure B to ANCIB minutes, 21 August 1945, Accession No. 2155, CBPC 27 , Cryptologic Archival Holding Area, NSA (TS).

8. John V. Connorton and Robert F. Packard, Secretariat, ANCICC, memorandum for the Subcommittee on Merger Planning (SMP), 30 August 1945, Accession No. 2468, CBIB 15, Cryptologic Archival Holding Area, NSA (TS).

9. John V. Connorton and Robert F. Packard, Secretariat. ANCICC, memorandum to members of ANCICC. "Recommendations of SMP concerning a location for the merged Army-Navy C.I. Organization," 7 September 1945, Accession No. 2468, CBIB 15, Cryptologic Archival Holding Area NSA(TS).

10. Ibid.

11. Brigadier General W. Preston Corderman, USA, Acting Chairman, ANCICC, memorandum for ANCIB, "Merging of Army-Navy Communication Intelligence Activities," 26 September 1945, Accession No. 2468, CBIB 15, Cryptologic Archival Holding Area, NSA (TS).

12. Ibid.

13. See Documentary Appendix to Part I, History of AFSA/NSA (Hereafter referred to as Documentary Appendix for AFSA/NSA) (Enclosure A), History Collection, NSA, Series V.A.(TS). Copies of all of the letters in the Eisenhower and Nimitz exchange of 1945 concerning merger are included in this file.

14. Fleet Admiral Ernest J. King, USN, memorandum for Secretary of the Navy, "The Navy's Interest in the Processing of Intercepted Foreign Communication." 14 October 1946, Accession No. 2468, CBIB 15, Cryptologic Archival Holding Area, NSA (TS).

15. See Documentary Appendix A for AFSA/NSA, Item 11.

16. See Documentary Appendix A for AFSA/NSA, Item 12.

17. Memorandum, STANCICC to STANCIB, "Coordination of the Army and Navy Conmunication Intelligence Activities," 15 February 1946 (Incl. A of minutes of STANCIB-STANCICC joint meeting, 15 February 1946), History Collection, NSA. Series XII.H. Box 36 (TSC).

18. Ibid.

19. Joseph N. Wenger, OP-20-G, memorandum for Captain W.K. Smedberg, III, (F-20), "Policy regarding collaboration between the Army and Navy in the Communication Intelligence Field," 30 May 1945, Accession No. 506lN, CBSB 23, Cryptologic Archival Holding Area, NSA (TS).

20. Minutes of 27th meeting of the State-Army-Navy Communications Intelligence Coordinating Committee,13 February 1948, History Collection, NSA. Serial XII.H, Box 37 (TS).

21. Minutes of joint meeting of the State-Army-Navy Communication Intelligence Board and the State-Army-Navy Intelligence Coordinating Committee, 15 February 1946, History Collection, NSA, Series XII.H. Box 36. (TSC).

22. Ibid.

23. British-United States Communication Intelligence Agreement, 5 March 1946, History Collection, NSA, Series XII.H. Box 47 (TSC).

24. See Wenger Collection in Cryptologic Archival Holding Area for information and chronology concerning the development of the separate appendices to the BRUSA Agreement. See Accession No. 9090 and 9091, CBIB 24.

25. Colonel W. Preston Corderman and Captain Joseph N. Wenger, memorandum for STANCIB, "Coordination of Army and Navy Communication Intelligence Activities," 5 April 1946, Cryptologic Archival Holding Area, Accession No. 2468, CBIB 15, NSA (TS). (Hereafter referred to as Corderman-Wenger Agreement.) The data for approval, 22 April 1946, is that given in Captain Wenger's memorandum to Chairman, UBCIB, "Report on the status and authority of USCIB-USCICC, its members and the Coordinator of Joint Operations," 7 February 1947. (Annex A, Item 19). See Accession No. 5061 N, Cryptologic Archival Holding Area. NSA (TS).

26. Ibid.

27. See Dr. George F. Howe's "Narrative History of AFSA/NSA," Part I, 28-30, History Collection, NSA, Series V.A. (TSC). (Hereafter cited as Howe's "History of AFSA/NSA").

28. Minutes of joint STANCIB/STANCICC meeting, 15 February 1946 (Item 3), History Collection, NSA, Series, XII.H. Box 36 (TS).

29. Howe's "History of AFSA/NSA," 28-30.

30. E.M. Huddleson, Jr., Director, Special Projects Staff, Department of State, memoranda for the Coordinator of Joint Operations, "Communications Intelligence Priorities," 24 June and 19 November 1946, Accession No. 1984, CBPC 27, Cryptologic Archival Holding Area, NSA (TSC).

31. Corderman-Wenger Agreement.

32. Memorandum for CJO, Subject: Report of Joint Intercept Control Group for the period 1 April 1946 to 28 February 1947,6 March 1947, from Lieutenant Colonel Morton A. Rubin, USA. Cryptologic Archival Holding Area, Accession No. 1379, CBPB 57, NSA (TS).

33. See enclosure to Corderman-Wenger Agreement of 5 April 1946, entitled "Coordination of the Army and Navy Communication Intelligence Activities, 15 February 1946," Accession No. 2468, CBIC 15, Cryptologic Archival Holding Area, NSA (S).

34. Memorandum from J. Edgar Hoover, Director. FBI, Lieutenant General Hoyt S. Vandenberg, Chairman, STANCIB, 14 February 1946, included as Annex Q to memorandum for members of USCICC, 25 August 1945, Accession No. 5061 N, CBIB 23, Cryptologic Archival Holding Area, NSA (Restricted).

35. Copy of memorandum from Lieutenant General Hoyt S. Vandenberg, USA, Chairman, STANCIB, to J. Edgar Hoover, Director, FBI. "FBI Representation on STANCIB," 4 June 1946, included as Annex S to memorandum for members of USBCICC, 25 August 1945, Accession No. 5061, CBIB 23, Cryptologic Archival Holding Area, NSA (Restricted).

36. Major General S.J. Chamberlin, Senior Member, USCIB, memorandum for Lieutenant General Hoyt S. Vandenberg, DCI, "Central Intelligence Group Representation on USCIB," 3 July 1946. For text of memorandum see J.N. Wenger memorandum for the Chairman, USCIB, "Report on Status and Authority of USCIB-USCICC." 7 February 1947, Annex U, Accession No. 5061, CBSB 23, Cryptologic Archival Holding Area, NSA (TS).

37. See Accession Nos. 1984 and 1377, Cryptologic Archival Holding Area, NSA, for a great variety of correspondence concerning the intelligence requirements of the civilian members of USCIB. These requirements are originated by the Department of State, CIA, and the FBI, and are addressed to the intelligence and priorities committees of USCIB and the CJO.

38. J.N. Wenger, memorandum for CJO, "Priority Problems," 24 October 1946, Accession No. 1984, CBSC 27, Cryptologic Archival Holding Area, NSA (TS). See also minutes of 34th meeting of USCICC (Item 3), 9 October 1946, History Collection. NSA, Series XII.H. Box 37 (TSC).

39. Grant Manson, State Department Member Of USCIB Intelligence Committee, "State Department Proposals for Modification of Requirements Expression," 25 February 1949, Accession No. 7674, CBSC 72, Cryptologic Archival Holding Area, NSA (TS).

40. George F. Howe Historical Study of COMINT under the Joint Operating Plan, 1946-1949, 164-65, History Collection, NSA, Series V.E (TSC).

41. Ibid.

42. Minutes of 30th meeting USCIB, 27 April 1948, with enclosure (UBCIB: 5/36, 9 April 1948), History Collection, NSA, Series XII.H. Box 37 (TSC).

43. Historical Study, Joint Operating Plan, 165.

44. Ibid., 164.

Chapter III

1. One year later, further changes took place at the national level, which primarily involved the military services and the defense establishment. Congress, in August 1949, amended the National Security Act converting the National Military Establishment into the Department of Defense, and making it a cabinet-level agency. By this time, there was a new secretary of defense, Louis A. Johnson. The new act made the three military services subordinate departments within the new Department of Defense structure. At the same time, Johnson acquired unqualified authority and control over the entire organization and became the "principal assistant to the President in all matters relating to the Department of Defense." Steven L. Rearden, *The History of the Office of the Secretary of Defense: The Formative Years, 1947-1950*, Vol. 1 (Washington, D.C.: Historical Office, Office of the SecDef, 1984), 53-60. (Hereafter cited as Rearden, *The Formative Years.*)

2. United States Communications Intelligence Board, Organizational Bulletin No. II, 31 July 1946, 8, History Collection. NSA., Series XII.H. Box 34 (TS). (Hereafter cited as USCIB Organizational Bulletin.)

3. Ibid., 6-7.

4. Ibid.

5. Ibid.

6. On 10 March 1945 General George C. Marshall, and Admiral Ernest J. King co-signed a Joint Memorandum to the Assistant Chief of Staff (G-2) and the Commanding General, Signal Security Agency and the Director of Naval Intelligence and the Director of Naval Communications, which formally established the Army-Navy Communications Intelligence Board. For the text of memorandum see the USCIB Organizational Bulletin, 2-3.

7. The term ULTRA is generally used to refer to the British exploitation of German messages encrypted on the Enigma machine. The term MAGIC refers broadly to the work of the United States in reading high-level diplomatic communications of the Japanese in the Pacific Theater, primarily on the Purple machine.

8. See Rearden, *The Formative Years*, 141.

9. Lay Report.

10. Thomas F. Troy. *Donovan and the CIA: A History of the Establishment of the Central Intelligence Agency* (Central Intelligence Agency, Center for the Study of Intelligence, 1981), 305-40. (Hereafter cited as Troy, *History of CIA*)

11. See Troy, *History of CIA*, 464, for the text of Truman memorandum to the Secretaries of State, War, and Navy.

12. Ibid., 351.

13. Ibid.

14. Ibid.

15. See Rearden, *The Formative Years*, 23-27.

16. Ibid., 24-27.

17. Ibid.

18. For the text of Truman memorandum establishing the Central Intelligence Group (CIG), see Troy, *History of CIA*, 464.

19. Howe's "History of AFSA/NSA," Part 1, 143.

20. Minutes of the 21st and 22d meetings of USCIB, 4 and 19 November 1947, History Collection, NSA, Series XII.H. Box 36 (TS).

21. See minutes of 25th and 27th meetings of the United States Communication Intelligence Coordinating Committee (USCICC), 17 February 1946 and 23 October 1946, 4-5 and 3-6, History Collection, NSA, Series XII.H. Box 36 (TS).

22. Starting with its 21st meeting on 4 November 1947, USCIB discussed the question of "'control" authorities over USCIB during many meetings. (Hereafter cited as minutes of 21st Meeting of USCIB.) In particular, see minutes for the 25th meeting of USCIB, 19 December 1947, History Collection, NSA, Series XIL.H. Box 36 (TSC).

23. See minutes of 21st meeting of USCIB.

24. W. Park Armstrong, Jr., Special Assistant to the Secretary for Research and Intelligence, memorandum to all the members of USCIB. "Amended Draft of NSCID Establishing USCIB," 7 June 1948, History Collection, NSA, Series XII.H. Box 34 (c). (Hereafter cited as Armstrong memorandum, 7 June 1948).

25. Minutes of 26th, 27th, and 28th meeting of USCIB, 6, 20 January 1948 and 3 February 1948, History Collection, NSA, Series XII.H. Box 38 (TS).

26. Lay Report, 32-34. See also minutes of 28th meeting of USCIB, 3 February 1948.

27. See minutes of 26th meeting of USCIB, 6 January 1948.

28. Ibid.

29. See minutes of 28th meeting of USCIB, 3 February 1948.

30. Ibid.

31. Ibid.

32. Forrestal memorandum for Secretaries of Army, Navy and Air Force, 13 February 1948, Spint File, CIA Archives, Langley, USCIB Charter (TS).

33. Grant Manson, State Department Liaison Officer, memorandum to Park Armstrong, Special Assistant. State Department, 3 March 1948, History Collection, NSA, Series XII.H. Box 47 (S)

34. Ibid.

35. Grant Manson, State Department Liaison Officer, memorandum for the record, "Current Developments on the Charter," 23 March 1948, History Collection, NSA, Series XII.H. Box 47 (TS).

36. R. H. Hillenkoetter, DCI, memorandum to Lieutenant General S.J. Chamberlin, Chairman, USCIB, 22 March 1948, History Collection, NSA, Series XII.H. Box 47 (S).

37. Lay Report, 34-35.

38. Lay Report, 34-36.

39. See minutes of 31st meeting of USCIB, 13 May 1948.

40. See Howe's "History of AFSA/NSA" for an excellent account of this entire sequence concerning the USCIB charter. In particular, see 122-29.

41. See Armstrong memorandum, 7 June 1948. The enclosure to the basic memorandum summarizes the for issues between CIA and other members of USCIB concerning the charter controversy.

42. Howe's "History of AFSA/NSA," 126-28.

43. Grant Manson, memorandum, "Various Views Concerning Developments on the Charter," 25 March 1948, History Collection, NSA, Series XII.H. Box 47 (TS).

44. See Armstrong memorandum of 7 June 1948.

45. C. T.R. Adams and J.E. Fitzpatrick, Ad Hoc Secretariat, memorandum for the DCI, "Action agreed upon at the time of the Ad Hoc meeting of the IAC held at 1410 on 11 June 1948," History Collection, NSA, Series XII.H. Box 34 (C). Admiral Roscoe H. Hillenkoetter, DCI, memorandum for the Executive Secretary of the National Security Council, "Proposed NSC Intelligence Directive, Communication Intelligence," 11 June 1948, History Collection, NSA, Series XII .H. Box 34 (S).

46. NSCID No.9, "Communications Intelligence," July 1948, History Collection, NSA, Series V.A. (C).

Chapter IV

1. Howe's "History of AFSA/NSA," Part I, 1-26.

2. Joseph N. Wenger, USN, memorandum for Director, AFSA, "Consolidation of CSA and ASA." 12 July 1949, Cryptologic Archival Holding Area, Accession No. 3791, CBPI 48, NSA (TS).

3. Interview of Stuart MacClintock, 25 February 1986, by Robert D. Farley, OH-05-88, Center for Cryptologic History, NSA (S).

4. Joseph N. Wenger, OP-20-G, memorandum for Civil Service Board, Staff and OINC NEGAT, Authority of Civil Service Appointees," 24 April 1946, Cryptologic Archival Holding Area, Accession No. 3791, CBPI 48, NSA (TS).

5. Joseph N. Wenger, memorandum for OP-20, "Civil Service," 18 December 1945, Cryptologic Archival Holding Area, Accession No. 3791, CBPI 48, NSA (TS).

6. Woodbury M. Burgess, USAF, "Briefing of Robertson Committee," 1957, History Collection, NSA, Series VI.C.

7. Kenneth C. Royall, Secretary of the Army, memorandum for James V. Forrestal, "United Armed Forces Security Agency," 24 July 1948. Accession No. 2459, CBID 14, Cryptologic Archival Holding Area, NSA (U).

8. Report of the Joint Committee of the Investigation of the Pearl Harbor Attack, U.S. Congress, 1946, 253, NSA Cryptologic Collection, NSA (U).

9. Howe's "History of AFSA/NSA," Part I, 144-46.

10. Rearden, *The Formative Years*, 53-56.

11. James V. Forrestal, memorandum to Secretaries of the Army, Navy, and Air Force, "Terms of Reference for the Committee on the Creation of a Unified Armed Forces Security Agency ," 19 August 1948, History Collection, NSA, Series VII. Box 4 (S) (Hereafter cited as Forrestal Letter.)

12. Ibid.

13. Minutes of the 1st meeting of the Committee on the Creation of a Unified Armed Forces Security Agency, 7 September 1948, History Collection, NSA, Series V.F (TS).

14. Forrestal Letter.

15. Earl E. Stone, memorandum for SecDef, "Committee on the Creation of a Unified Armed Forces Security Agency–Report of," 30 December 1948, Historical Collection, NSA, Series V.F (TS). (Hereafter cited as Stone Board Report).

16. Ibid., (Tab D–Army View).

17 .Ibid., 6.

18. Ibid., 6.

19. Ibid., 5.

20. Ibid., (Part A), 9.

21. Minutes of 41st meeting of USCIB, 17 June 1949 (USCIB 25/1), "Establishment of Armed Forces Security Agency " History Collection, NSA, Series XII.H. Box 37 (TSC). (Hereafter cited as 41st meeting of USCIB.)

22. Roscoe H. Hillenkoetter, DCI, memorandum for SeeDef, "Report of the Committee on the Creation of a Unified Armed Forces Security Agency," 14 January 1949, History Collection, NSA, Serisa XII.H. Box 38 (TSC).

23. 41st meeting of USCIB, and Howe's "History of AFSA/NSA," 155.

24. Brownell Committee, Report to the Secretary of State and SecDef, 13 June 1952, p. 21, History Collection, NSA, Seriea V.F (TSC). (Hereafter cited as Brownell Report.)

25. Minutes of 13th meeting of AFSAC, 28 April 1960, (AFSAC 25/74), 14-15, History Collection, NSA, Series XII.A (TS).

26. George F. Howe, "The Early History of NSA," *Cryptologic Spectrum*, (Spring 1974), 11-17.

27. Howe's History of AFSA/NSA, 158-59.

28. Rearden, *The Formative Years*, 410-22.

29. Captain Paul R. Schratz, USN (Ret.), "The Admiral's Revolt," U.S. Naval Institute Proceedings (February 1986).

30. Interview of Brigadier General Carter W. Clarke, 3 May 1983, by Robert D. Farley, OH-03-83, History Collection, NSA. (Hereafter cited as Carter Clarke interview.)

31. Howe's "History of AFSA/NSA," 59.

32. Louis Johnson, memorandum for the Joint Chiefs of Staff, "Organization of Cryptologic Activities within the National Military Establishment," 20 May 1949, Accession No. 6490, CBIB 44, Cryptologic Archival Holding Area (TS). (Hereafter cited as Johnson Directive Establishing AFSA)

33. Louis Johnson, letter to Secretary of State, 20 May 1949, Accession No. 6490, CBIB 44, Cryptologic Archival Holding Area, NSA (TS).

34. Johnson Directive establishing AFSA.

35. 41st meeting of USCIB.

36. Howe's "History of AFSA/NSA," 165-66.

37. Minutes of 41st meeting of USCIB.

38. Johnson Directive establishing AFSA.

39. JCS 2010/6, Charter for Armed Forces Communication. Intelligence Advisory Council, 28 July 1949, Cryptologic Archival Holding Area, Accession No. 1404, CBSD 11, NSA.(TS). (Hereafter cited as JCS 2010/6.)

40. JCS 2010/12, Change in title of "Armed Forces Comunications Intelligence Advisory Council," Cryptologic Archival Holding Area, Accession No. 1404, CBDD 11, NSA (TS).

41. While Stone reported as Director on 15 July 1949, the earlier date of Johnson's Directive, 20 May 1949, is generally cited as the date for the establishment of AFSA.

42. Howe's "History of AFSA/NSA," Part II, 3-6.

43. Ibid.

44. AFSAC 25/29, Progress Report as of 30 January 1950 by Director, AFSA, to AFSAC, 27 January 1950, Accession No. 5253, CBSB 71, Cryptologic Archival Holding Area. NSA (TS) and Howe's "History of AFSA/NSA," Part II, 15-45. Howe's "History of AFSA/NSA," Part II, 15.

46. Ibid. Part II, 40.

47. Travis M. Hetherington, Chief Plans and Policy Divisions, memorandam for General Canine, "Establishment of a USCIB Coordinator's Group," 13 May 1952, with enclosure,. 3-4 of enclosure, Cryptologic Archival Holding Area, Accession No. 2960, CBSB 33, NSA (TS).

48. Ibid.

49. Earl Stone, memorandum for members of AFSAC, Progress Reports of 30 January 1950," 27 January 1950, Cryptologic Archival Holding Area, Accession No. 5253, NSA (TS).

50. Johnson Directive establishing AFSA.

51. AFSA/Air Force Agreement on Task Assignment to AFSS Mobile Intercept Sites, 22 September 1950 (AFSAC: 60/24), Cryptologic Archival Holding Area, Accession No. 4820, CBSB 57, NSA (TS).

52. Joseph N. Wenger, memorandum to George F. Howe, NSA Historian. Comments on Historical Study, 25 May 1960, Part D, XII, History Collection, NSA, Series VI.D (TS). (Hereafter cited as Wenger, comments on AFSA/NSA History.)

53. Howe's "History of AFSA/NSA," 78-30, 92-94.

54. JCS 2010/23, "Division of Responsibility between Armed Forces Security Agency and the Military Services," 18 November 1950,Cryptologic Archival Holding Area, Accession No. 1404, CBDD 11, NSA.(TS).

55. Ibid.

56. During 1949 and 1950, CIA and State continually stressed the issue of their lack of an authoritative role in USCIB activities. In particular, see minutes of 41st meeting of USCIB, 17 June 1949, 2-8, History Collection, NSA, Series XII.H. Box 34 (TS).

57. Louis W. Tordella, "Presentation to Eachus Study Group," 10 September 1968, 12, History Collection, NSA, Series VI.D (TSC).

58. Rear Admiral Earl E. Stone, Director, AFSA, memorandum for Assistant Chief or Staff, G-2, Director of Naval Intelligence, Director of Intelligence, USAF, "Formulation of Special Intelligence Requirements," 18 August 1950, Cryptologic Archival Holding Area, Accession No. 39058, NSA. (TS).

59. JCS 2010/28, "Succession of Directors, Armed Forces Security Agency," 17 February 1951, Cryptologic Archival Holding Area, Accession No. 1404, CBDD 11, NSA. (TS).

60. Ibid., and Howe's "History of AFSA/NSA", Part II, 137.

61. Howe's "History of AFSA/NSA," Part II, 137-38.

62. Interview of Rear Admiral Earl E. Stone, 9 February 1983, by Robert D. Farley, OH-03-83, Center for Cryptologic History, NSA and Wenger, comments AFSA/NSA History, Center for Cryptoogic History, NSA.

63. Carter Clarke interview.

64. Major General Ralph J. Canine, Director, NSA, Presentation to Agency seniors at Arlington Hall Station, 25 November 1952, Cryptologic Archival Holding Area, Accession No. 4918, CBNB 61, NSA (TS).

Chapter V

1. See minutes for 48th meeting of USCIB, concerning CONSIDO, 13 January 1950. Accession No. 6490 (Folder #2), Cryptologic Archival Holding Area, NSA (TSC).

2. Howe's "History of AFSA/NSA," Part II, 126-29.

3. Louis Johnson, memorandum for the Joint Chiefs of Staff, "Maximum Exploitation of COMINT," 19 May 1949, with enclosure, Accession No. 6491, CBIB 45, Cryptologic Archival Holding Area, NSA (TSC). (Hereafter cited as Johnson Memorandum, 19 May 1949, concerning CONSIDO.)

4. Ibid.

5. Major General S. Leroy Irwin, USA, Director of Intelligence, memorandum for Joint Intelligence Committee, "Conference with General McNarney on Draft. CONSIDO Paper," 20 October 1949, Accession No. 6491, CBIB 45, Cryptologic Archival Holding Area, NSA (TS).

6. Johnson memorandum, 19 May 1949, concerning CONSIDO.

7. USCIB 26/1, Item 7 of the Agenda for 47th meeting of USCIB, 2 December 1949, "Presentation of Draft CONSIDO Paper," Accession No. 6490 (Folder 12), CBIB 45, Cryptologic Archival Holding Area, NSA (TS).

8. USCIB 26/4 Item 2, Section I, of the Agenda for the 48th meeting of USCIB, held 13 January 1950, "Draft Proposal on the Establishment of CONSIDO," Accession No. 6490 (Folder #1), CBIB 44, CBIB, History Collection, NSA (TSC). (Hereafter cited as USCIB 26/4, 13 January 1950.)

9. USCIB 26/4, 13 January 1950.

10. Rear Admiral R.H. Hillenkoetter, USN, DCI, memorandum for USCIB Members, 12 January 1950, Accession No. 6491, CBIB 45, Cryptologic Archival Holding Area. NSA. (TSC).

11. USCIB 26/4, 13 January 1950.

12. Ibid.

13. USCIB 26/7, Item 2 of the Agenda for the 49th meeting of USCIB, held on 16 March 1950, "Draft Proposal on Establishment of CONSIDO," History Collection, NSA, Series XII.H. Box 38.1 (TSC).

14. USCIB 26/13, Item 2 of the Agenda for the 53rd meeting of USCIB, held on 14 July 1950, "Draft Proposal on Establishment of CONSIDO," History Collection, NSA, Series XII.H. Box 38.1 (TS.)

15. See minutes of 30th meeting of USCIB, 27 April 1948, 2-7, w/encl (memorandum for Rear Admiral Roscoe H. Hillenkoetter, Director, Central Intelligence, for Lieutenant General S.J. Chamberlin. Rear Admiral T .B. Inglis, and Rear Admiral Earl E. Stone, dated 21 April 1948. History Collection, NSA, Series XII.H, Box 38.1 (TSC)

16. Charles P. Collins, CIA, Chairman, USCIB Intelligence Committee, memorandum to Coordinator of Joint Operations, "Recurrent lntelligence Requirements List No. 3." 15 May 1950, Accession No. 1378, CBPS 57, Cryptologic Archival Holding Area, NSA (TSC).

17. Ibid.

18. Captain Mason, Chief of Office Operations, teletype message for Captain Wenger, V/DIR, undated, but apparently originated soon after the start of the Korean War. During this period, there existed a teletype circuit (known as "pony" circuit) between the Office of Operations at Arlington Hall Station and the AFSA Directorate at Naval Security Station. The circuit was used primarily for exchange of operational communication between AFSA officials. Copy of message is in Wenger Collection. See Accession No. 9139, CBIB 26, Cryptologic Archival Holding Area, NSA (TSC). (Hereafter cited as Mason to Wenger Message.)

19. The U.S. COMINT Effort During the Korean Conflict, June 1950 - August 1963, unpublished manusript, 8 January 1954), 1-2, History Collection, NSA, Series V.M (TSC). (Hereafter Korea.)

20. Ibid., 2.

21. Ibid., 3.

22. Mason to Wenger Message.

23. Korea, 2-3.

24. Rear Admiral Earl E. Stone, Director, AFSA, memorandum for Assistant Chief of Staff, G-2, Director of Naval Intelligence, Director of Intelligence, USAF, "Formulation of Special Intelligence Requirements," 18 August 1950, Accession No. 39058A, Cryptologic Archival Holding Area, NSA (TS). See also responses from Army, Navy, and Air Force as promulgated by AFSAC 60/25, Accession No. 39058A, NSA (TS).

25. Korea, 33-34, 57-58.

26. Louis Johnson, memorandum for the Joint Chiefs of Staff, "Organization of Cryptologic Activities within the National Military Establishment," 20 May 1949, Accession No. 1404, CBDD 11, Cryptologic Archival Holding Area, NSA (TS).

27. Brigadier General W.N. Gilmore, Chief, ASA, memorandum for Major General Bolling, Assistant Chief of Staff, G-2, "Korean Intercept Effort Prior to the Outbreak of Korean Hostilities," 5 October 1950, Accession No. 9139, CBIB 26, Cryptologic Archival Holding Area, NSA (TSC).

28. AFSAC 60/26: Report by the Director, AFSA, to the Joint Chiefs of Staff, via the Armed Forces Security Agency Council, "Division of Responsibility Between AFSA and the Services," 18 September 1950, Accession No. 5864 (Folder #3), CBIB, Cryptologic Archival Holding Area, NSA (TS).

29. Ibid. See also Joint memorandum from Rear Admiral Felix Johnson, Director of Naval Intelligence, Rear Admiral John R. Redman, Director of Naval Communications, and Captain L.S. Howeth, Head Security Branch, Naval Communications, for Chairman, Armed Forces Security Agency Council, "Comments on AFSAC 60/26," 27 October 1950, Accession No. 5864 (Folder #3), CBIB 31, Cryptologic Archival Holding Area, NSA (TS).

30. Stone wrote these informal remarks on the Army's response to AFSAC 60/30, 19 October 1950, Accession No. 9092, CBIB 24, Cryptologic Archival Holding Area, NSA (TS).

31. AFSAC 60/30, 19 October 1950: Major General C.P. Cabell, Director of Intelligence, USA, memorandum for Chairman, AFSAC, "Report by DIRAFSA, AFSAC 60/26," Accession No. 5864 (Folder #3), CBIB 31,

Wenger Collection, Cryptologic Archival Holding Area, NSA (TS).

32. AFSAC 60/33, 24 October 1950: Major General Alexander R. Bolling, Assistant Chief of Staff, G-2 memorandum for Chairman, AFSAC, "Comments on AFSAC 60/26," Accession No. 9092, CBIB 24, Cryptologic Archival Holding Area, NSA (TS).

33. Memorandum for the Record by Captain Joseph N. Wenger, Deputy Director, AFSA, "Comments on AFSAC 60/26," November 1960, Accession No. 5864, (Folder #3) CBIB 31, Wenger Collection, Cryptologic Archival Holding Area, NSA (TS).

34. Captain Joseph N. Wenger, USN, Deputy Director, AFSA, memorandum for Director, AFSA, "Army Comments on AFSAC 60/26," 26 October 1950, Accession No. 5864, CBIB 31, Cryptologic Archival Holding Area, NSA (TS). (Hereafter cited as Wenger memorandum, 26 October 1950).

35. Wenger memorandum, 26 October 1950. See also memorandum from Captain Rufus L. Taylor, OP-32, "Comments on JCS 2010/24," 15 January 1951, Accession No. 5864, CBIB 31, Cryptologic Archival Holding Area, NSA (TS).

36. Wenger memorandum, 26 October 1950.

37. Admiral Earl E. Stone, Director, AFSA, memorandum for Major General Alexander R. Bolling, USA, "AFSAC 60/26," 1 November 1950, Accession No. 9092, CBIB 31, Cryptologic Archival Holding Area, NSA (TS).

38. AFSAC 60/42: Admiral Earl E. Stone, Director, AFSA, memorandum for Members of AFSAC, "Division of Responsibility Between AFSA and the Military Services," 24 November 1950, Accession No. 6864, CBIB 31, Cryptologic Archival Holding Area, NSA (TS).

39. Minutes of 53rd meeting of USCIB, 14 July 1950, 6-11 (USCIB 77/4: Selective Intelligence Mobilization), History Collection, NSA. Series XII.H. Box 36.1 (TS).

40. Consumer Liaison Units, 1949-1957, April 1957, 7-10. Accession No. 10684, CBRI 52, Cryptologic Archival Holding Area, NSA (TSC). (Hereafter cited as Consumer Liaison)

41. Ibid., 2-11

42. Ibid., 3-11.

43. Colonel Roy H. Lynn, memorandum for Coordinator of Joint Operations, "Scheduling of USCIB Committee Meetings," 2 May 1949, Accession No. 1426, CBPB 63, Cryptologic Archival Holding Area NSA (S). Colonel

David Wade, USAF, Vice Commander, memorandum for CJO, "Transmission of Classified Material to this Headquarters," 9 May 1949, Accession No. 1426, CBPB 83, Cryptologic Archival Holding Area, NSA (S).

44. Consumer Liaison, 19-23.

45. Ibid.

46. Within the State Department, the Special Projects Staff (SPS) became responsible for the receipt and dissemination of COMINT. In May 1947, Robert F. Packard served as the first COMINT liaison officer for the SPS, which was then located at the Pentagon. See Consumer Liaison, 33-34.

47. Minutes of 53rd meeting of USCIB, 14 July 1960, 6-11, (USCIB 77/4.: Selective Mobilization), History Collection, NSA, Series XII.H, Box 38.1 (TS).

48. T. Achilles Polyzoides, Director, Special Projects Staff, Department of Stete, memorandum for Director, AFSA, "Establishment of a Department of State (Special Projects Staff) Liaison Unit at AFSA 02," 25 June 1951, History Collection, NSA, Series XII.H (TS). Admiral Stone, Director, AFSA, memorandum for the Special Assistant, Department of State, "'Establishment of a Department of State (Special Project. Staff) Liaison Unit at AFSA 02," 5 July 1951, History Collection, NSA, Series XII.H (TS).

49. See minutes of 30th meeting of USCIB, 27 April 1948, 2-7, w/encl (memorandum for Rear Admiral Roscoe H. Hillenkoetter, Director, Central Intelligence, for Lieutenant General S.J. Chamberlin, Rear Admiral T.D. Inglis, and Rear Admiral Earl E. Stone, dated 21 April 1948. History Collection, NSA Series XII.H, Box 38 (TSC).

50. Ibid.

51. Minutes of 41st meeting of USCIB, 17 June 1949 (USCIB 25/1) "Establishment of Armed Forces Security Agency," History Collection, NSA, Series XII.H. Box 38.1 (TSC)

52. AFSA/Air Force Agreement on Task Assignments to AFSS Mobile Intercept Sites, 22 September 1950 (AFSAC 60/24), Cryptologic Archival Holding Area, Accession No. 4820, CBSB 57, NSA (TS). Wenger "Comments on Historical Study."

53. Major General Charles A. Willoughby, G-2, Far East Command. message dated 12/1908Z March 1951, addressed to ACSI, Washington, Accession No. 9139, Cryptologic Archival Holding Area, NSA (TS). Captain

J.S. Holtwick, USN, Chief, Office of Operations, memorandum for B.K. Buffham and H. Conley, "Visit to FEC," 12 March 1951, Accession No. 9139, Cryptologic Archival Holding Area, NSA (9). Major General Charles A. Willoughby, Assistant Chief of Staff, G-2, Far East Command, memorandum for Herbert Conley and Benson K. Buffham, 15 March 1951, Accession No. 9139, Cryptologic Archival Holding Area, NSA (TSC). Report of Tour of Far East Command, by B.K. Buffham and H.L. Conley, 4 April 1951, Accession No. 9139, Cryptologic Archival Holding Area, NSA (TSC),

54. Brownell Committee, Report to the Secretary of State and SecDef, 13 June 1952. 62, History Collection, NSA, Series V.F . (TSC). (Hereafter cited as Brownell Report.)

55. Ibid. See also Captain Redfield Mason, Chief, AFSA 02, memorandum for Commanding General, ASA, "Korean Linguists for ASAPAC," 1 August 1950, Accession No. 9139, Cryptologic Archival Holding Area, NSA (TSC).

56. These improvements are generally attributed to the intervention of Soviet advisors who became alarmed over the communications procedures of the North Koreans. See Richard Chun, "A Bit on the Korean COMINT Effort," (unpublished manuscript, 1971), 3, History Collection, NSA, Series V.M. (TSC).

57. Robert E. Drake, "The COMINT Role in the Korean War," (unpublished manuscript, Circa 1954), 6, History Collection, NSA, Series V.M (TSC).

58. Louis W. Tordella, "Presentation to Eachus Study Group," 10 September 1968, 12, History Collection, NSA, Series VLD (TSC).

58. The minutes of the USCIB meetings from 1949-1950 provide evidence of the beginning of a gradual change in the attitude of the CIA representatives. DCI Hillenkoetter, who served as the third DCI from 1 May 1947-7 October 1950, became more assertive in his efforts to carve out a greater role for CIA in the national intelligence process. In particular, see USCIB 26/4, 13 January 1950. See also the minutes of the 41st meeting of USCIB, concerning the establishment of AFSA, 17 June 1949. History Collection, NSA, Series XII.H. Box 38.1 (TSC). (Hereafter cited as minutes of 41st meeting of USCIB, 17 June 1949.)

Chapter VI

1. During the USCIB meetings of 1949 and 1950, CIA and State continually raised the issues of their lack of an authoritative voice in guiding the activities of AFSA. In particular, see the minutes of the 41st meeting of USCIB, 17 June 1949, 2-8, History Collection, NSA, Series X.II.H (TS). (Hereafter cited as minutes of 41st meeting of USCIB, 17 June 1949.)

2. In later years, a portion of the CONSIDO concept was resurrected anew and implemented by the SIGINT Policy Board. Twenty-five years after the disapproval of the CONSIDO Proposal, the National Foreign Intelligence Board established the National SIGINT Requirements System (NSRS) as a community requirements mechanism in 1975.

3. Minutes of 41st meeting of USCIB, 17 June 1949.

4. The standing requirement for a unanimous vote on major decisions had prevailed in USCIB for a number of years. Consequently, it became very difficult for the representatives of CIA and State to win support for a favorable decision in USCIB on controversial issues, particularly when they sought to acquire an enhanced role in influencing the AFSA intelligence process. See NSCID No. 9, "Communications Intelligence," 1 July 1948, History Collection, NSA, Series V.A. (TS).

5. John Ranelagh, *The Agency: The Rise and Decline of the CIA*. New York: Simon and Schuster, Inc., 1986, 102-11 and Rearden, *The Formative Years*, 20-27.

6. Lawrence K. White on the Directors. Oral History Interview by Dino Brugioni and Urban Linehan, *Studies in Intelligence* (CIA), Winter 1987. Vol. 31, No.4. 1-6.

7. Walter B. Smith, DCI, memorandum for Executive Secretary. National Security Council, "Proposed Survey of Communications Intelligence Activities," 10 December 1951 (CIA Archives) (TSC).

8. Robert A. Lovett and Dean G. Acheson letter to Brownell, Bohlen, Magruder and Jackson, 28 December 1951 (Exhibit A in Brownell Report). History Collection, NSA, Series V.F (TSC).

9. Ibid.

10. Lieutenant General Ralph J. Canine, "Memorandum for the Record," 7 February 1952, Cryptologic Archival Holding Area. Accession No. 6009, CBSD 75. NSA.(S).

11. Ibid.

12. Ibid.

13. Brownell Report, 62.

14. Ibid., 50.

15. Ibid., 6.

16. Ibid., 134.

17. Ibid., 6, 121-28.

18. Ibid., 121-28.

19. Ibid., 7 of cover letter.

20. Ibid., 124-28.

21. Ibid., 129-34.

22. Ibid., 131-34.

23. Ibid., 53-56.

24. Ibid.

25. Ralph J. Canine, Director, AFSA, memorandum for SecDef, "Brownell Committee Report," 8 July 1952. Cryptologic Archive Holding Area, Accession No. 5741 N, CBSC 33, NSA (TS).

26. Memorandum for the Record by Rear Admiral J.N. Wenger, Deputy Director for Intelligence, AFSA, "Draft of Proposed Presidential Directive for Organizing the COMINT Activities of the U.S.," 14 October 1952, Accession No. 5741 N, CBSC 33. Cryptologic Archival Holding Area, NSA (TS).

27. Harry S. Truman, memorandum for the Secretary of State and the Secretary of Defense, 24 October 1952. History Collection. NSA, Serial V.A (TS).

28. National Security Council Intelligence Directive No.9, Revised, Communications Intelligence, 24 October 1952, Cryptologic Archival Holding Area. Accession Number 14842, CBBD 76, NSA (TS).

29. Ibid.

30. Harry S. Truman, memorandum for the Secretary of State and the SecDef, "Communications Security (COMSEC) Matters," History Collection, NSA, Series V.A (TS).

31. Robert A. Lovett, SecDef, memorandum. for the Service Secretaries, Joint Chiefs of Staff. Director, National Security Agency, "Interim Implementation of NSCID No.9," Revised 4 November 1952, History Collection. NSA, Series V.A (TS).

32. Ibid.

33. Lieutenant General C.D. Cabell, Director Joint Staff, memorandum for General Bradley, General Vandenberg, General Collins, and Admiral Fechteler, "National Security Agency," 9 February 1963, Cryptologic Archival Holding Area, Accession No. 2466, CBIB 15, NSA (S).

34. Ibid.

35. Almost four decades later, the national directives unequivocally identify NSA as the central authority in the field of U.S. cryptology. NSCID No. 6, issued in February 1972 charges the director with "full control" over all SIGINT collection and processing activities, which now include COMINT and electronic intelligence (ELINT) processing. But, as in 1952, there still exist a number of grey areas between NSA and the services, mainly because of charter limitations and exceptions to the authority of NSA. Moreover, national directives also provide for the delegation of SIGINT authorities, as necessary, to the military services for conducting a number of direct support activities, including tactical ELINT operations. While considerable progress has been made in pulling together U.S. SIGINT activities into the semblance of a single system, divided control still exists.

Notes on Sources

Most of the documents used in the preparation of this history are in the holdings of the NSA/CSS Records Center and Archives and the Center for Cryptologic History. A diverse number of sources originated these holdings, which reflect a broad range of departmental, national, and operational relationships extending over this period of cryptologic history. The NSA/CSS Records Center and Archives holds three basic groups of resource materials: the stored records that are held for a temporary period pending a disposition review by the owning organization; the retired records that are undergoing an appraisal to determine their archival value; and the accessioned records that are filed as permanent Agency records in the Cryptologic Archival Holding Area (CAHA), or Archives. The Agency's Center for Cryptologic History maintains its own research collections.

Among Agency records, two separately organized collections deserve special mention. These are the accessioned records of the Archives and the special collections maintained by the Center for Cryptologic History. For the researcher, there is very little distinction between the kind of records in the Archives and those maintained in the Center for Cryptologic History. There are major distinctions, however, in the method of organization and arrangement of the documents for retrieval purposes – and in the continuity accorded historical themes. These differences in organization stem mainly from basic differences in the approach to records keeping as well as factors associated with the organizational evolution of each organization. In the course of developing into today's structures, each organization underwent a different sequence of growth, and each developed its own operating concepts and methodology.

Starting with the AFSA period and extending into the NSA years, the Office of the Adjutant General (AG) served as the first administrator of cryptologic records. During the early 1950s the AG established the beginnings of the Agency's records management program and directed the creation of a Records Repository for the retention of vital records. These early rudimentary actions safeguarded from destruction massive holdings acquired from the World War II era as well as other essential records associated with the establishment of AFSA and NSA. However, as NSA directed its primary energies toward its operational missions and as organizational changes occurred, the position of Agency officials concerning the priority of non-operational tasks of this nature became clear. The resource allocation officials consistently demonstrated little enthusiasm for the program and generally provided only token support in terms of resources and priorities.

In responding to its new national role, NSA commenced a pattern of frequent organizational change that extended from 1952 until the late 1970s. This pattern of recurring institutional change impacted unfavorably on the direction and emphasis accorded its records management function. During this period of approximately twenty-five years, the management responsibility for the task rotated among at least six key components: Office of the Adjutant General, Office of Administrative Services, Comptroller, Office of Policy, Office of Management Services, and Office of Telecommunications. Despite this cycle of change and the continuance of the strictures on resources, the records management program achieved some progress over the years. But, overall, these circumstances clearly affected the quality and scope of the program and impeded its implementation. They also encouraged the creation of special collections of historians and history-minded technicians.

As the Agency expanded, the lack of storage facilities emerged as another major problem in the conduct of the Agency's records management program. Until the late 1970s, the Agency resorted to storing its record holdings at several different locations within the Agency as well as a number of locations outside NSA's control. These external locations included facilities at Crane, Indiana; Arlington Hall Station; Vint Hill Farms; Torpedo Station, Alexandria, Virginia; and Fort Holabird, Baltimore. The dispersal of records represented an inefficient method of operation and impacted negatively on the various steps in the review and disposition cycle. As a practical matter, however, perhaps the greatest damage occurred in the retrieval process. The dispersal of stored, retired, and permanent records not only affected the quality and timeliness of service provided to operational elements, but it also impeded the retrieval efforts of researchers and historians.

All of these factors contributed to a lack of direction and stability for the entire records program. The big push for change did not come until 1977, when the Agency established the Cryptologic Archival Holding Area (CAHA) – or Archives, as it is commonly called. The establishment of an Archives stemmed from action by President Carter directing the mandatory declassification of intelligence documents that were thirty years old or older. In complying with Executive Order 12065, Admiral Inman, the director of NSA, ordered that a new urgency be placed on declassification matters and on the records management program of NSA. As an integral part of this action, Inman directed the immediate establishment of a new archival office to assume archival responsibility for all elements of the Agency and to function under the control of the Director's Policy Staff. By 1980, with the physical relocation of all of the stored, retired, and accessioned records in the Office of Archives and Repository Services, the Agency concluded its first serious attempt to establish an archival program.

Of the records processed thus far by the archives since its establishment in 1977, the accessioned records generally start with the World War I era and extend to 1960. The accessioned records yielded significant information for this study. These records are arranged and filed under a nine-letter code group, called a Cryptologic Record Group (CRG), which identifies the file location as well as the "origin, geographic pertinence, and subject content" of the record. The most useful part of this immense collection is its subject correspondence file of letters, memoranda, reports, and other correspondence between the Army and Navy cryptologic organizations and between the military services and officials in the defense establishment, the National Security Council, the White House, and other executive departments. There are also a number of Special Collections within these holdings, such as the Wenger Collection, personal papers, and various project files, that proved to be extremely valuable.

The Center for Cryptologic History traces its origins to the AFSA era when a history office was established as a very small element within the training division (AFSA

14). But shortly after the establishment of NSA in 1952, the history function received new attention and emphasis at the Directorate level. This change occurred mainly because of the interest of General Canine, who wanted the events associated with the establishment of NSA to be documented from a historical viewpoint. He supported the recruitment of three professional historians and created a new history element, which functioned as a part of his Plans and Policy Staff. Following the appointment of Dr. George F. Howe as the first professional historian, the history program enjoyed some degree of high-level support and recognition which lasted for several years. Within a few years, however, following the retirement of General Canine, the political situation began to change for the history element. By the mid-1950s, as the growing Agency became increasingly preoccupied with more pressing operational considerations, the position of the history group began to deteriorate. Lacking any sponsorship at the Directorate level, the group encountered great difficulty in obtaining support and resources for conducting even a modest history effort.

Like the early Records Office, the History Office became a casualty of frequent reorganization actions, as well as periodic resubordinations within the Agency. By 1989 the History Office had been placed organizationally in at least seven different Key Components. These Key Components included the Training Office, Plans and Policy, the Central Reference Organization in the Office of Operations, the Policy Staff, the Management Organization, the National Cryptologic School, and the Office of Telecommunications.

Vice Admiral Studeman's 17 September 1989 memorandum announced the decision to establish the Center for Cryptologic History as an element of the director's staff and marked a new departure for the Agency's history program. Personnel were transferred from the Office of Telecommunications the following month to constitute the core of the new organization, with Henry F. Schorreck retaining his position as NSA historian. Components of the Center include the oral history program and the publishing arm (which also publishes *Cryptologic Quarterly*, the Agency's professional journal) as well as the research collections and historians.

During the years of austerity, the NSA Historian of necessity performed only limited research activities and concentrated on the development of historical records relating to the evolution of the cryptologic structure of the United States. Today, the Center for Cryptologic History, by virtue of its History Collection, is a major holder and authority on early U.S. cryptologic records. The collection is designed to assist in meeting the needs of Agency researchers and in providing information support services to Agency officials.

The records in the History Collection begin with the American Revolution and extend into the present. These holdings are divided into Series, generally on a chronological basis, with further subdivisions made topically. The History Collection contains published and unpublished manuscripts, a broad range of policy and operational correspondence, personal collections, crisis files, historical studies, and transcripts of oral history interviews. Another important collection of historical records exists as a totally separate entity within the Center. The "Cryptologic Collection" became a part of the History Collection in 1987. Originally a part of the Technical Documents Section of the NSA Library, the Cryptologic Collection is slanted heavily toward technical matters. It contains a wealth of material on cryptographic systems from the World War II period and earlier, descriptions of cryptanalytic solutions and techniques, and extremely useful information concerning cryptologic organizations.

Among the records acquired from the three services for this early period, the holdings contributed by the Army are noteworthy and are the most prolific of the three military services. During the early 1980s, the U.S. Army Intelligence and Security Command (INSCOM) welcomed NSA personnel in their screening of INSCOM holdings from World War I to the post-World War II period. As a result of this cooperation, the Archives now holds an immense collection of Army cryptologic records for the early years of cryptology.

The archival holdings acquired from the Navy and Air Force vary considerably. With particular reference to the Navy, the Archives does hold a collection of Army-Navy records that devolved to NSA following the organization-al realignments occurring during the postwar years. Initially this sequence began with the Joint Army-Navy Operating Agreement of 1946, which forced each service to move toward closer cooperation on cryptologic matters. The Agreement resulted in a first-time consolidation of a broad range of Army-Navy operating documents and correspondence, including the records of the various Joint Committees. As a part of this process, the Navy also intermixed with its contemporary holdings some earlier naval records dating to the immediate prewar period. As further organizational changes occurred, this collection of joint Army-Navy holdings passed to the custody of each successor structure. The cycle of institutional change, extending over six years, included the establishment of AFSA in 1949 and ended with the creation of NSA in 1952. Today this legacy of Army-Navy records is invaluable, not only for research, but also as a source of accurate perspective on the nature and problems of these early joint operations.

Because of its later arrival as a third cryptologic service, the Air Force records for this early period do not start until after World War II. The Air Force created the Air Force Security Service (AFSS) as a separate command in Texas in 1949, five months before the establishment of the Armed Forces Security Agency. During a transition period, the initial AFSS structure relied on the Army Security Agency for administrative and operational support. By the time of the Korean War, however, the AFSS started to function as an independent service and acquired its own facilities and targets. Still, even for this latter period, there is a paucity of internal Air Force documentation at NSA concerning the inner workings of the new cryptologic service. This may have been due, in part, to the remote location of AFSS Headquarters in Texas. Today the Air Force unit histories appear to constitute the bulk of the Air Force cryptologic records in the archives. But for the purposes of this report, these unit histories proved to be of minimal value.

The combined holdings of the Archives and the Center for Cryptologic History contain significant documentation issued not only by the military services (both the cryptologic and intelligence organizations), but also by the evolving United States Communication Intelligence Board (ANCIB–STANCIB–USCIB) and the civilian consumer agencies. In particular, the correspondence, agendas, and minutes of the initial policy boards (ANCIB–STANCIB–USCIB) provided exceptional perspective about the nature of the conflicts and power struggles taking place within the intelligence structure during the postwar period from 1945 to 1952. The internal correspondence from some consumer agencies gave special insights into the unity reflected by the representatives of CIA and State in their joint opposition to the exclusive military control of the COMINT effort. This block of records also provided enlightening perspective, once again from the point of view of the non-military consumer, about the AFSA structure and the activities of the Brownell Committee.

Oral interviews, conducted mainly by Robert D. Farley, the first NSA Oral Historian, and his successors and selected Special Research Histories (SRH), resulting from declassification, helped fill out the documentary record.

Primary Sources

U.S. Congress

U.S. Congress, House Hearings, Committee on Expenditures in the Executive Departments, National Security Act of 1947, 80th Congress, 1st Session, 1947.

U.S. Congress, Senate Hearings, *National Security Act Amendments of 1949*, 81st Congress, 1st Session, 1949.

U.S. Congress, Joint Committees, *Report of the Joint Committee of the Investigation of the Pearl Harbor Attack*, 79th Congress, 2nd Session, 1946. (Joint Committee Print.)

Presidential Papers

Acheson, Dean, Henry L. Stimson, and James V. Forrestal, Memorandum for President Truman, "Continuation and Extension of Collaboration with the British in the Communications Intelligence Field," undated, NSA Center for Cryptologic History.

Roosevelt, Franklin D., memorandum for the Director of the Budget, 8 July 1942, NSA Archives.

Roosevelt, Franklin D., memorandum for General Marshall, 9 July 1942, NSA Center for Cryptologic History.

Truman, Harry S., memorandum for the Secretaries of State, War, and Navy, 12 September 1945, NSA Center tor for Cryptologic History.

Truman, Harry S., memorandum for the Secretary of State and the Secretary of Defense, 24 October 1952, NSA Center for Cryptologic History.

Truman, Harry S., memorandum for the Secretary of State and the Secretary of Defense, "Communications Security (COMSEC) Matters," 24 October 1952, NSA Center for Cryptologic History.

National Directives

United States National Security Council, NSCID Number 9, *Communications Intelligence*, 1 July 1948.

United States National Security Council, NSCID Number 9, Revised, *Communications Intelligence*, 24 October 1952.

International and United States Military Agreements

"Army-Navy Agreement concerning Allocation of Diplomatic Traffic," 30 June 1942, NSA Center for Cryptologic History.

"Agreement between British Code and Cipher School and U.S. War Department concerning Special Intelligence," 10 June 1943, NSA Archives.

Marshall, George C. and Edward J. King, "Joint Army-Navy Agreement for the Exchange of Communications Intelligence," 4 February 1944, NSA Center for Cryptologic History.

"Outline of the Collaboration in Cryptanalysis between the Army and the Navy," 18 August 1944, NSA Archives.

"British-United States Communications Intelligence Agreement," 5 March 1946, NSA Center for Cryptologic History.

"Corderman-Wenger Agreement concerning Coordination of Army and Navy COMINT Activities," 6 April 1946, NSA Archives.

"Establishment of a United States Combined Intelligence Liaison Center in Great Britain," 3 May 1946, NSA Archives.

"AFSA/Air Force Agreement on Task Assignments to AFSS Mobile Intercept Sites," 22 September 1960, AFSAC 60/24, NSA Archives.

Official Documents Issued by United States Intelligence Boards

Most important to this study were the minutes, agendas, and organizational bulletins of the evolving ANCIB-STANCIB-USCIB structures. High-ranking officials of the military cryptologic organizations and the civilian agencies presented their views on the COMINT structure, its functioning, and its placement in the national intelligence structure. From 1945 to 1952, the issuances of the early intelligence boards revealed the conflicts and struggles of the intelligence community relating to the issues of unification and the consolidation of cryptologic responsibilities. These unique sources are in a single consolidated grouping within the NSA Center for Cryptologic History. This brief list reflects only a sampling of the documents available for this period

Minutes of the Army-Navy Communications Intelligence Coordinating Committee, 1944- 1945

Minutes of the Army-Navy Communications Intelligence Board, 1945-1946

Minutes of the State-Army-Navy Communications Intelligence Board, 1945

Minutes of STANCIB concerning the establishment and operation of the Joint Operating Plan, 1946-1949

Minutes of USCIB Meetings concerning the establishment of AFSA in 1949 and the parallel proposal concerning the CONSIDO Plan

Executive Department Documents and Reports

Army-Navy-FBI Allocation Committee, "Report of Conference Appointed to Study Allocation of Cryptanalysis," 30 June 1942, NSA Archives.

"Report of Joint Army-Navy Intercept Control Group for the period 1 April 1946 to 28 February 1947," 6 March 1947, NSA Archives.

Stone Board Report to the Secretary of Defense concerning the Creation of a Unified Armed Forces Security Agency, 30 December 1948, NSA Center for Cryptologic History.

Progress Report by Director, AFSA, to Armed Forces Security Advisory Council, 27 January 1950, NSA Archives.

Report by Director, AFSA, to Armed Forces Security Advisory Council, concerning Division of Responsibility Between AFSA and the Services, 18 September 1950, AFSAC 60/26, NSA Archives.

Joint Military Memorandum to Army and Navy COMINT Committee, 31 July 1940.

Joint Army and Navy Report to Chief Signal Officer and the Director of Naval Communications, 3 October 1940.

Report by Secretariat of Army-Navy Communications Intelligence Coordinating Committee concerning Merger Planning, 30 August 1945.

United States Communications Intelligence Board Organizational Bulletin No. II, 31 July 1946.

Secretary of Defense, "Promulgation of Terms of Reference for the Committee on the Creation of a Unified Armed Forces Security Agency," 19 August 1948.

JCS 2010/12, Organizational Announcement concerning Armed Forces Communications Intelligence Advisory Council, 28 July 1949.

Secretary of Defense Directive, "The Establishment of the Armed Forces Security Agency within the National Military Establishment," by Louis A. Johnson, 20 May 1949.

JCS 2010/23, "Division of Responsibility Between AFSA and the Military Services," 18 November 1950.
AFSAC 60/49, "Announcement of the Establishment of AFSS Group Headquarters, 18 June 1951.

Secretary of Defense, "Implementation of NSCID Number 9," Revised, by Robert A. Lovett, 4 November 1952.

Historical Studies and Monographs

History of the Signal Security Agency in World War II, "Organization, 1929-1945" (ASA 1945), Vol. I, Part I.

History of the Signal Security Agency. "The Japanese Army Problems–Cryptanalysis." (ASA 1947).

Parker, Frederick D., *A New View to Pearl Harbor: United States Communication Intelligence, 1924-1941*, United States Cryptologic History Series. (NSA 1988)

Oral Interviews

Clarke, Carter W., Interviewed by Robert D. Farley, Clearwater, Florida, 3 May 1983, NSA Center for Cryptologic History.

MacClintock, Stuart, Interviewed by Robert D. Farley, Fort Meade, Maryland, 25 February 1986, NSA Center for Cryptologic History.

Stone, Earl E., Interviewed by Robert D. Farley, Carmel, California, 9 February 1983, NSA Center for Cryptologic History.

White, Lawrence K., Interviewed by Dino Brugioni and Urban Linehan, Washington, D.C., 8 June 1972, CIA History Office.

Special Research History (SRH)

SRH200 – *Army-Navy Collaboration*, 1831-1945, Part 2.

SRH230 – *The Role of COMINT in the Battle of Midway*, Henry F. Schorreck.

Unpublished Studies

Chun, Richard A., A Bit on the Korean Effort. Typescript. Working note prepared for NSA History Office, 1971.

Drake, Robert E., The COMINT Role in the Korean War. Typescript. Study prepared for the Director, NSA, circa 1954.

Howe, George F., "Narrative History of AFSA/NSA," Parts I-IV, (NSA 1959).

Howe, George F., "Historical Study of COMINT under the Joint Operating Plan, 1946-1949" (NSA).

The U.S. COMINT Effort During the Korean War June 1950-August 1953. Typescript. 1954

Lay, James S., History of USCIB. Vol. I. Typescript. Study prepared for CIA Historical Office, undated.

Secondary Sources

Hough, Richard, *The Greatest Crusade–Roosevelt, Churchill, and the Naval Wars* (New York: William Morrow and Co., 1986).

Ranelagh, John, *The Agency: The Rise and Decline of the CIA* (New York: Simon and Schuster, Inc., 1986).

Rearden, Steven L. *The History of the Office of the Secretary of Defense, The Formative Years, 1947-1950* (Washington, D.C.: Government Printing Office, 1984).

Troy, Thomas F., *Donovan and the CIA: A History of the Establishment of the Central Intelligence Agency* (Central Intelligence Agency, 1981).

Schratz, Paul R., *The Admiral's Revolt* (U .S. Naval Institute Proceedings, February 1986).

INDEX